W0111236

ADVANCES IN BIOCHEMICAL ENGINEERING

Volume 10

Editors: T. K. Ghose, A. Fiechter, N. Blakebrough

Managing Editor: A. Fiechter

With 48 Figures

Springer-Verlag Berlin Heidelberg GmbH 1978

ISBN 978-3-662-15466-3   ISBN 978-3-540-35682-0 (eBook)
DOI 10.1007/978-3-540-35682-0

© by Springer-Verlag Berlin Heidelberg 1978
Library of Congress Catalog Card Number 72-152360
Originally published by Springer-Verlag Berlin Heidelberg New York in 1978.
Softcover reprint of the hardcover 1st edition 1978

2152/3140-543210

Editorial

The appearance of volume 10 of the series "Advances in Biochemical Engineering" is a good reason to state that this rapidly developing speciality has overcome its struggle against traditional boundaries and has established itself as a discipline in its own right. As in other such developments, the dynamics of specialization and reintegration of sciences follow their own laws. This joint venture, essentially based on convergent developments in microbiology and biochemistry on the one hand and technology and engineering on the other, has proved to be very successful in theory and rewarding in practice. The topics treated and evaluated in the review articles of this series show convincingly that Biochemical Engineering can significantly contribute to the analysis and solution of a considerable variety of relevant problems.

There are many possibilities and opportunities for Biochemical Engineering to make further contribut ions to material progress and to stimulate significant developments. Just a few of them, which may serve as examples, will be mentioned here.

First, the improvement of methods enabling the continuous growth of any kind of microorganism under optimal conditions will require safe and simple ways of regulation. The use of such procedures in large scale production largely depends on energy input, yield and quality of the product, but no less on their reproducibility.

Second, the search for new microbes and for mutants with novel features also falls within the scope of Biochemical Engineering. Taking into consideration that bacteria and molds are experienced masters in biological warfare, it is not unreasonable to assume that there exists a number of metabolic products still awaiting detection and chemical and biological characterization. Third, it will be of considerable practical interest to widen the scope of available tools in Biochemical Engineering, e.g. by further investigating the use of purified enzymes and enzyme combinations, whether free or carrier-bound, for any kind of synthetic or degradative procedures.

Obviously the most important problem of our time is to provide enough food for mankind. Since the danger persists that the world population will continue to grow more rapidly than food production, more efficient ways of production are urgently needed. If there still is some hope that science can significantly contribute to the solution of this problem, then it is expected to come in the first place from genetics and biochemical engineering.

It has happened twice in the history of mankind that fundamental achievements drastically changed the living conditions and the food production capacity on this planet: first, when animals were domesticated and dairying was started; second when systematic agriculture with its increasing monocultures permitted harvesting of huge crops. Is it unreasonable to propose that taming of suitable microbes with all the consequences regarding production capacity may ultimately lead to an analogous improvement in man's food supply?

Even if one is more modest and does not anticipate a real scaling up in productivity of basic foodstuff such protein, there remain many other possible developments which may help to improve the nutritional situation. It is an established fact that the large scale synthesis of nucleotides, amino-acids (notably glutamate), certain polysaccharides, vitamins and other food components or additives is more readily accomplished by microbial processes today than by the classical chemical methods used in the past. In this context it is of interest to note that those countries with a leading position in the world market are also those who first became aware of the great impact of the astonishingly high and versatile synthetic capacity of microbes in low cost mass production.

An analogous situation may apply with regard to measures aimed at the protection of our environment. Microbes have always played an important role in biodegradation. The more frequent the accidents leading to drastic water pollution and the heavier the load of substances resistant to recycling, the more microbial strains and techniques which can efficiently counteract pollution are needed. Selective binding of toxic heavy metals such as lead, mercury or cadmium, oxidative degradation of hydrocarbons, detergents and pesticides are just some examples showing what might be achieved in this respect.

Knowledge in Biochemical Engineering is advancing rapidly. In view of its considerable prospective significance this newly formed discipline deserves a high degree of priority. This not only means that scientists active in this field should be supported and encouraged; it also means that the exchange of views should be promoted, as by this series of progress reports.

However, it also implies that those young biochemists, microbiologists and engineers who feel responsible and concerned should get an opportunity to collaborate. Finally, although Biochemical Engineering has become an independent discipline, it should never be forgotten that a close interdisciplinary collaboration and a special effort to understand and to integrate the partner's point of view have greatly contributed to the rapid progress made in this field.

H. Aebi

Contents

Design and Operation of Immobilized Enzyme Reactors   1
W. H. Pitcher, Jr., Corning, New York (USA)

Biotechnology of Immobilized Multienzyme Systems   27
S. A. Barker, P. J. Somers, Birmingham (Great Britain)

Carriers for Immobilized Biologically Active Systems   51
R. A. Messing, Corning, New York (USA)

Industrial Applications of Immobilized Biocatalysts   75
P. Brodelius, San Diego, California (USA)

Starch Hydrolysis by Immobilized Enzymes.   131
Industrial Applications
B. Solomon, Rehovot (Israel)

Design and Operation of Immobilized Enzyme Reactors

Wayne H. Pitcher, Jr.
Corning Glass Works, Corning, NY 14830, USA

Contents

1   Introduction   1
2   Reactors   2
3   Reaction Kinetics and Reactor Performance   5
4   Mass Transfer and Related Effects   9
    4.1 Internal Mass Transfer   9
    4.2 External Mass Transfer   13
    4.3 Transient Behavior of Immobilized Enzyme Reactors   14
    4.4 Electrostatic Effects   15
    4.5 Backmixing   15
5   Heat Transfer   16
6   Temperature Effects   17
7   Immobilized Enzyme Activity Loss   17
8   Operating Strategy   17
9   Multienzyme Systems   20
10 System Cost Estimates   20
11 General Design Considerations   21
12 Future Trends   22
13 Table of Symbols   22
14 References   23

This review encompasses recent advances in the design and operation of immobilized enzyme reactors for industrial applications. Basic immobilized enzyme reactor engineering concepts are described as a reference point for recent innovations. Although practical examples are cited, the subject is approached from the viewpoint of reactor design and operation and the potential general applicability of new concepts or developments. Areas reviewed include reactor types, reactor performance, operating strategy, and general design and economic considererations. Most of the progress reported in this field is in the form of refinements, rather than basic innovations.

1 Introduction

The design and operation of immobilized enzyme reactors is perhaps the key element in any processing scheme utilizing immobilized enzymes. With this in mind, it sometimes is surprising to realize that the immobilized enzyme reactor may, in fact, be only a small part of the total process. Certainly the immobilized glucose isomerase reactors contributed only in a minor way to the cost and plant area necessary for the production of high fructose corn syrup. Furthermore, the design of immobilized enzyme reactors has relied heavily on conventional process design concepts and thus has not changed radically since the first commercialization.

It is the purpose of this chapter to examine recent developments in the design and operation of immobilized enzyme reactors in the context of traditional process design concepts. It will be seen that major innovations and improvements are infrequent, but that subtleties can often be important in a field still in its infancy. Often generalizations are difficult with each system requiring in-depth study. Thus practical examples play an important role in illustrating the various aspects of reactor design and operation. Specific topics such as reactors, reaction kinetics, mass transfer, and operating strategy are treated individually with concluding sections concerning system cost estimates and general design considerations to put the individual subjects in perspective. It should be noted that this author [1, 2] and Vieth *et al.* [3] have previously attempted comprehensive reviews of the engineering of immobilized enzyme systems.

2 Reactors

Immobilized enzyme reactors can be separated into several different categories including batch reactors, continuous stirred tank reactors, fixed-bed reactors, and fluidized-bed reactors. These reactor types can also be combined or modified. Addition a recycle loop can, for example, be used to change the behavior of a fixed-bed reactor system. Extensive literature concerning reactor design for heterogeneous catalysis exists and can be readily applied to many immobilized enzyme systems.

Perhaps the simplest reactor is the batch reactor. Unless the immobilized enzyme can readily be recovered for re-use after the reaction has been carried to completion, this type of reactor fails to take advantage of one of the primary assets of immobilized enzymes, long life or enzyme stability. Usually only in small scale applications, such as for high value low volume products in the pharmaceutical industry, will the simplicity of operation outweigh the higher labor costs of batch versus continuous operation.

There are a variety of methods for recovering immobilized enzymes from batch operations. For some applications the enzyme could be recovered by ultrafiltration even if it were not immobilized. O'Neill *et al.* [4] attached enzymes to soluble, high molecular weight polymers, which allowed separation by ultrafiltration even in the presence of relatively high molecular weight reaction products. In the same vein, Coughlin *et al.* [5] prepared a soluble derivative by coupling NAD to alginic acid using 1,2,7,8-diepoxyoctane. This derivative could be precipitated by lowering the pH below 3. Simple filtration or centrifugation of immobilized enzyme particles can also be used. Robinson *et al.* [6] immobilized enzymes on magnetizable particles, which could then be recovered magnetically using existing technology.

Other approaches have included a stirred tank reactor in which the immobilized enzyme was enclosed in mesh containers (panels) attached to a stirrer to give agitation without immobilized enzyme attrition [7]. Of course reaction media can be recycled through continuous flow immobilized enzyme reactors to give batch reaction conditions.

Snam Progetti [8–10], an Italian company, has chosen to exploit its unique process of entrapping enzymes within the pores of wet spun synthetic fibers (such as cellulose triacetate or γ-methyl-polyglutamate) by using recycle reactors.

Many of these techniques can also be used to retain the immobilized enzyme in continuous stirred tank reactors. In fact, several of the preceding examples were originally used in continuous stirred tank reactors. Closset *et al.* [11, 12] analyzed and operated tubular membrane reactors for the hydrolysis of starch by β-amylase. Starch and enzyme were contained by the membrane, which retained the starch but was permeable to the maltose product.

The type of reactor used almost exclusively in commercial scale operations, at least those known to the public, is the fixed-bed. Early examples were Clinton Corn Processing Company using immobilized glucose isomerase for conversion of glucose to fructose and Tanabe Seiyaku Company with aminoacylase bound to DEAE-Sephadex for resolution of D,L-amino acids. Due to its high efficiency, ease of operation and general simplicity, this reactor type will probably continue to dominate large-scale commercial applications. There are certain problems unique to some forms of fixed-bed reactors. In packed beds of particulate enzyme derivatives, a trade-off between small particles with low diffusional resistance and causing high pressure drops and larger particles resulting in diffusion limited reaction rates, but lower pressure drop. Reactor variants such as the monolith, developed as the support for automatic exhaust catalysts, or the rolled membrane and backing systems of Emery [13] and Venkatasubramanian and Vieth [14] have a more open structure with higher void volume. This structure can be an advantage in the case of high specific enzyme activity where the short diffusional distances will result in higher effectiveness factors than for particulate derivatives of practical size. For expensive enzymes, a high effectiveness factor may be essential for acceptable process economics. Particulate material in the feed is also less likely to plug this more open reactor design. However, most commercial or potential commercial applications utilize enzyme composites that have high effectiveness factors with acceptable particle size. In this case, activity per unit volume will actually be higher than for the open reactor designs. Again the simplicity and potential low cost of the particulate packed bed are attractive. For inorganic particles that can be regenerated for re-use, the carrier cost may in some cases be negligible.

Tubular reactors with enzymatically active walls are extreme examples of open structure fixed-bed reactors. In analytical applications where efficiency is not important or in biomedical applications such as extracorporeal shunts where low pressure drop and low turbulence are necessary, tubular reactors may be used.

Horvath *et al.* [15] have studied plug flow, the introduction of inert gas along with the feed to a tubular reactor. They found that the secondary flow patterns within the liquid plugs increased transport to the tube wall for the case of laminar flow in small-diameter tubes.

Another type of reactor, the fluidized bed, has received some attention recently [16]. Its advantages include low pressure drop (and thus low pumping costs) and the ability to handle fine particulate feeds. The velocities necessary for fluidization may result in residence times insufficient to achieve the desired conversion. Solutions to this problem might be recycle or the use of a series of fluidized beds. According to Charles *et al.* [17]

and Lee *et al.* [18], who used immobilized lactase and glucose isomerase, respectively, under certain conditions, fluidized-bed reactors give performance identical to fixed-bed reactors. The extent of bed expansion can evidently influence the performance of fluidized-bed reactors. Emery and Revel-Chion [19] compared experimental data with predictions from a modified dispersion model for fluidized-bed immobilized enzyme reactors. Shumate [20] reported work with a conically tapered (large end up) fluidized bed which exhibited improved stability. Of course, merely enlarging the diameter of the top of the reactor will also decrease the likelihood of washing out the carrier particles. Gelf and Boudrant [21] described the use of magnetizable particles as enzyme supports. In their system, a magnetic collar around the top of the column prevented the particles from leaving the top of the fluidized-bed reactor even at extremely high flow rates. Gas evolution or contacting can also be handled more readily in a fluidized-bed than in a packed-bed reactor.

The choice of reactor design depends on the requirements of the particular immobilized enzyme system in question. The high efficiency of the packed-bed reactor, which frequently approaches plug flow performance, is an advantage hard to overcome. Plug flow and batch reactors are more efficient than ideal continuous stirred-tank reactors (backmix reactors) when reaction kinetics are higher than zero in order. Enzyme catalyzed reactions frequently follow Michaelis-Menten kinetics, between zero and first order in substrate concentration dependence. The ratio of ideal plug flow to back-mix reactor volume necessary to achieve the same conversion level as a function of conversion for several values of So/Km (So = initial substrate concentration and Km = Michaelis constant) is shown in Fig. 1. The advantage of the plug flow versus backmix reactor is even greater in the case where product inhibition exists. There are instances,

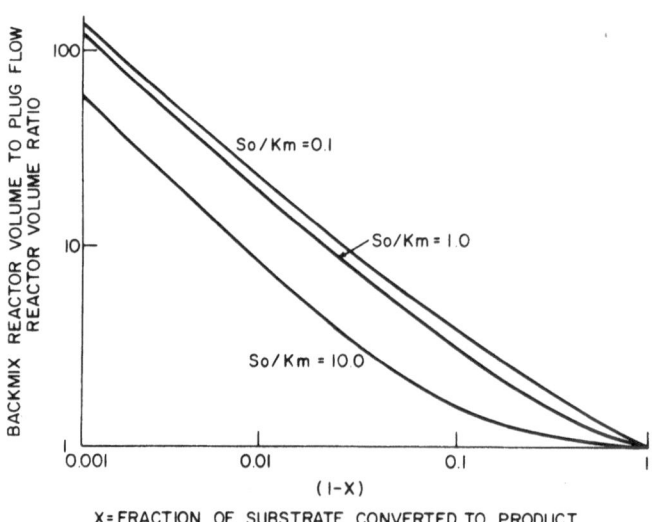

Fig. 1. Plug flow vs backmix reactor performance

such as substrate inhibition, where backmix reactors are superior, but these are few in practice. The fixed-bed reactor also usually results in the highest enzyme loading per unit reactor volume.

Somewhat related to the questions of efficiency and enzyme loading is the problem of mass transfer limitation. Although in many practical cases, the packed bed is at no disadvantage, the other types of reactors can make use of smaller immobilized enzyme particles or thin membranes or otherwise avoid diffusion limitations.

Stirred tank reactors, batch or continuous, and fluidized beds subject the immobilized enzymes to harsher treatment than do fixed-bed reactors. While the dangers of enzyme deactivation as a result of shear have probably been exaggerated [22], the enzyme support material can be severely affected.

Control of pH and temperature can readily be achieved in stirred tank reactors. Recycle reactors and fluidized beds can also readily be maintained at the desired temperature. However, this advantage is largely illusory in the case of temperature control since most enzyme catalyzed reactions have low heats of reaction. Operational stability is perhaps easiest to achieve with fixed-bed reactors.

The packed bed is more susceptible to plugging and channeling than many other fixed-bed reactors, as well as other reactor types.

Energy cost will not generally prove a major factor in reactor choice. Labor costs, of course, favor the continuous reactors.

Additional discussions of design considerations will be reviewed in the final section of this chapter.

3 Reaction Kinetics and Reactor Performance

Knowledge of reaction kinetics and the mode of reactor operation can be used to predict the behavior of ideal immobilized enzyme reactors. Normally three basic types of ideal reactors may be encountered: batch reactors (well-stirred), continuous stirred tank or backmix reactors, and plug flow reactors. The performance of real, non-ideal, continuous reactors generally falls somewhere in between that of the backmix and ideal plug flow reactors.

The following treatment of kinetics uses simple, irreversible Michaelis-Menten type kinetics as an example, but the same mathematical manipulations are appropriate for other types of enzyme kinetics. The reaction velocity for the irreversible Michaelis-Menten type kinetics can be written as

$$v = \frac{kES}{K_m + S},\tag{1}$$

where $v = -V_s(dS/dt)$ (V_s is substrate solution volume and t is time), k is the turnover number (a constant), E is the amount of enzyme, S is the substrate (reactant) concentration, and K_m is the Michaelis constant. Equation (1) can be integrated from time zero to time t to obtain the following expression for a batch reaction

$$kEt/V_s = S_0X - K_m \ln(1 - X),\tag{2}$$

where S_0 is initial substrate concentration and $X = (So - S)/So$. For an ideal plug flow reactor where each volume element of fluid proceeds through the reactor behaving as an infinitesimal batch reactor, not mixing with the adjacent fluid elements, Equation (2) can also be written as

$$KE/F = S_0 X - K_m \ln (1 - X),\qquad(3)$$

where F is the volumetric flow rate.
For a backmix reactor, a simple material balance gives

$$v = F(S_0 - S).\qquad(4)$$

where S_0 is the concentration of the substrate entering the reactor and S is the concentration in the reactor and at the outlet. The right hand sides of Eqs. (1) and (4) can then be set equal to each other.

$$F(S_0 - S) = \frac{kES}{K_m + S}.\qquad(5)$$

This equation can be rearranged and the fractional conversion term X substituted to give the final equation.

$$\frac{kE}{F} = X \left(\frac{K_m}{1 - X} + S_0 \right).\qquad(6)$$

Equivalent equations for reversible Michaelis-Menten substrate inhibited and product inhibited kinetics are given in Table 1.
Generally integrated rate equations or reactor operating data can be expressed graphically as a conversion versus residence time curve. In the case of heterogeneous catalysis, such as immobilized enzyme catalysis, the residence time should be normalized to reflect the amount of catalyst (enzyme) present. Normalized residence time can be expressed in several ways. For a batch reactor, it can be expressed as

$$\frac{Wt}{V_s} \text{ or } \frac{E_t t}{V_s},$$

where W = weight of immobilized enzyme and E_t = total enzyme activity. For a continuous reactor, normalized residence can be written as

$$\frac{W}{F} \text{ or } \frac{E_t}{F}.$$

An example comparing continuous and batch reactor data for the same immobilized enzyme [23] is shown in Fig. 2.
For ideal batch and plug flow reactors with no external mass transfer limitations, the conversion versus normalized residence time curves will coincide. These types of curves, or equations describing the curves, can be determined experimentally and then used to calculate enzyme activity from conversion and flow rate measurements. To be specific,

Table 1. Integrated rate equations

Kinetic type and rate equation
Michaelis-Menten

$$v = \frac{kES}{K_m + S}$$

Plug flow: $\dfrac{kE}{F} = S_0 X - K_m \ln(1 - X)$

Backmix: $\dfrac{kE}{F} = X\left\{\dfrac{K_m}{1 - X} + S_0\right\}$

Reversible Michaelis-Menten

$$v = \frac{kE\,(S - P/K)}{K_m + S - K_m P/K_p}$$

Plug flow: $\dfrac{kE}{F} = X_e S_t \left\{(1 - K_m/K_p)(X_t - X_i) + \dfrac{K_m}{S_t} + 1 - X_e \dfrac{K_m K_e}{K_p} \ln \dfrac{X_e - X_i}{X_e - X_t}\right\}$

Backmix: $\dfrac{kE}{F} = \dfrac{(X_t - X_i)(K_m + S_t - X_t S_t + K_m S_t X_t/K_p)}{1 - X - X_t/K}$

Substrate Inhibition

$$v = \frac{kE}{1 + K_m/S + S/K'_m}$$

Plug flow: $\dfrac{kE}{F} = S_0 X - K_m \ln(1 - X) + \dfrac{S_0^2 X}{K'_m} - \dfrac{S^2 X^2}{2\,K'_m}$

Backmix: $\dfrac{kE}{F} = X S_0 \left\{1 + \dfrac{K_m}{S_0(1 - X)} + \dfrac{S_0(1 - X)}{K'_m}\right\}$

Competitive product inhibition

$$v = \frac{kES}{S + K_m(1 + P/K_i)}$$

Plug flow: $\dfrac{kE}{F} = S_0(1 - K_m/K_i)(X_t - X_i) - (K_m + \dfrac{K_m S_t}{K_i}) \ln (\dfrac{1 - X_e}{1 - X_i})$

Backmix: $\dfrac{kE}{F} = \dfrac{(X_t - X_i)\left\{S_t(1 - X_t) + \dfrac{K_m + K_m X_t S_t}{K_i}\right\}}{1 - X_t}$

---

if this type of curve has been determined, the observed conversion level can be used to find the corresponding normalized residence time. From this value and the substrate flow rate (or substrate volume and elapsed reaction time), the enzyme activity at the time of the observation can be calculated.

Although this pragmatic approach is often necessary, it can also be valuable to gain an understanding of the enzyme kinetics to allow predictions using equations of the type given in Table 1. A number of authors [24–27] have treated the subject of enzyme kinetics comprehensively. However, some recent developments and practical considerations are worth noting.

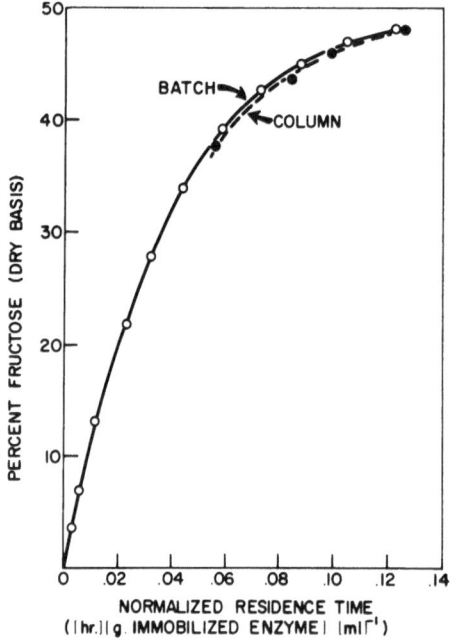

Fig. 2. Conversion vs normalized residence time

Traditionally, enzyme kineticists have used linear plots of initial rate data to determine the constants in the initial rate expression. Three types of plots have been used. The most common is the Lineweaver-Burk plot, $1/V$ versus $1/S$. The Hofstee plot, v versus v/S, and the Eadie or Hanes plot, S/V versus S are used much less frequently. Recently several authors have proposed using full time course kinetic studies in conjunction with the integrated rate equation to determine kinetic constants [28–30]. In many practical cases, once the reaction has been studied over its full time course, the resulting plot is of more use than the calculated constants. However, having an accurate kinetic model may be valuable when the effect of changing conditions must be predicted. For commercial applications, the immobilized enzyme systems are normally studied under conditions simulating actual operation. Conversion levels are usually relatively high. Eisenthal and Cornish-Bowden [31, 32] have raised a number of practical and statistical objections to the use of the three plots based on linear transformations of Eq. (1). Perhaps of greater significance, they proposed a simple new graphical procedure for estimating enzyme kinetic constants [31]. In the simplified example in Figure 3, the reaction velocity point on the vertical axis is connected with the substrate concentration point on the horizontal axis for each rate measurement. Each intersection point is an estimate of K_m and V_m (defined as KE). The best estimates, $\hat{K}_m$ and $\hat{V}_m$, are taken as the medians from each set of estimates. Among other advantages, this method introduces less bias in the case of outliers.

Yun and Suelter [33] have attempted to meet some of Cornish-Bowden's [34] objections to the use of the integrated equation to determine K_m and V. Their approach does account for different kinetic models including product inhibition, meeting one of

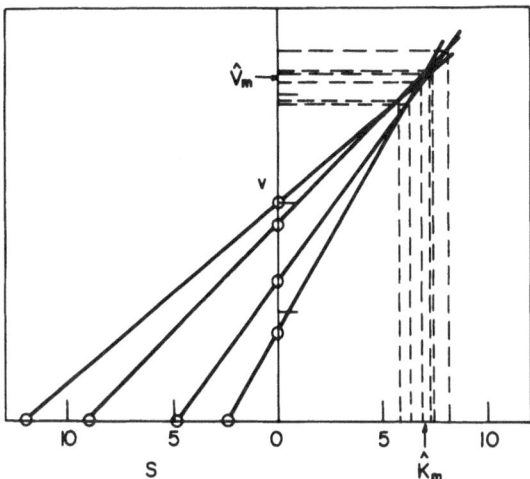

Fig. 3. Kinetic constant estimation

Cornish-Bowden's criticisms. However, the problem of enzyme inactivation during assay is more difficult to handle. In cases where the enzyme is relatively stable, the procedure may be of interest.

Increasing appreciation for statistically sound methods of kinetic analysis is evidenced by several recent articles. Storer *et al.* [35] found that variances increased with initial velocities of enzyme catalyzed for three experimental systems, contrary to assumptions made in the past. Fjellstedt and Schlesselman [36] have proposed a method for statistical analysis of initial velocity and/or inhibition data. Duggleby and Morrison [37] used non-linear regression techniques to analyze time course reaction data.

One factor to remember in applying any of these techniques of treating kinetic data to immobilized enzymes is that, although the soluble enzyme kinetics may be modeled simply, mass transfer limitations may change the apparent kinetics so as to make these types of simple plots useless. However, with proper allowances for the implications of scale-ups, the conversion versus normalized residence time plots should always give an accurate description of reaction behavior.

4 Mass Transfer and Related Effects

4.1 Internal Mass Transfer

A large body of information exists concerning the general problem of simultaneous reaction and diffusion. Probably the most comprehensive treatment and review of this subject is Satterfield's [38] book. Even the literature concerned with pore diffusion specifically in immobilized enzyme system is extensive [39–50] and somewhat repetitive in that many similar approaches have been reported. The coverage of the subject here will consist of a brief review of the basic approach to the problem, some recent studies, and practical experimental ways of evaluating internal diffusional effects. When a reaction is promoted by an enzyme immobilized within a porous material, a substrate concentration gradient is established inside the pores, with concentration decreasing with the distance from the porous body surface. For most immobilized enzyme systems with pore diffusion limitations, it is reasonable to assume that steady-state conditions exist. However, Vieth et al. [3] reviewed the available literature concerned with transient behavior of immobilized enzyme systems. Shyam et al. [51] have carried out a detailed analysis of transient behavior in bed of porous spherical immobilized enzyme composites.

The differential equation describing steady-state diffusion with chemical reaction in a sphere is

$$\frac{d^2S}{dr^2} + \frac{2}{r}\frac{dS}{dr} = \frac{v_i}{D_{eff}}, \tag{7}$$

where r is the radial distance in the sphere, v_i is the intrinsic reaction rate (usually a function of substrate concentration), and D_{eff} is effective diffusivity. This equation can be solved analytically only when $v_i = k_v S^m$, where m is a constant. Cases such as Michaelis-Menten kinetics must be solved numerically.

The ratio of observed or apparent reaction rate, to the rate if no diffusion limitation existed, called the effectiveness factor, η, can be calculated from the solution to Eq. (7). This effectiveness factor will normally be less than unity for immobilized enzyme systems.

A number of generalized plots have been developed which relate the effectiveness factor to various types of general moduli. These moduli fall into two general categories, those that depend on knowledge of intrinsic kinetics and those that utilize observed reaction rate information.

Bischoff [52] expressed η graphically as a function of a general modulus, M, which can be applied to Michaelis-Menten type kinetics assuming D_{eff} is constant at various levels of S/K_m. This modulus, which involves intrinsic kinetic constants, for flat plate geometry is

$$M = L \left(\frac{V_m}{2\,K_m D_{eff}}\right)^{1/2} \left(\frac{S}{K_m + S}\right) \left(\frac{S}{K_m} - \ln\left(1 + S/K_m\right)\right)^{-1/2}, \tag{8}$$

where for M greater than 2, $\eta = 1/M$. The plot for several levels of S/K_m is shown in Fig. 4.

An example of the same type of plot [1, 2, 38] for a modulus depending on observed reaction rates is shown in Fig. 5. This modulus,

$$\Phi_L = \frac{L^2}{D_{eff}} \left(\frac{1}{V_c}\frac{dn}{dt}\right)\frac{1}{S}, \tag{9}$$

which can be modified for spherical geometry, assumes D_{eff} to be constant, where $D_{eff} = D\Theta/\tau$, Θ = porosity (void fraction), τ = tortuosity (ratio of actual diffusion path to straight line distance), and L = flat plate thickness. This type of modulus is particularly useful in practice since the term $1/V_c\,dn/dt$ is readily measurable, simply being the observed reaction rate per unit volume of porous carrier. Moo-Young and Kobayashi [53] have solved numerically the more complex cases for product and substrate inhibition.

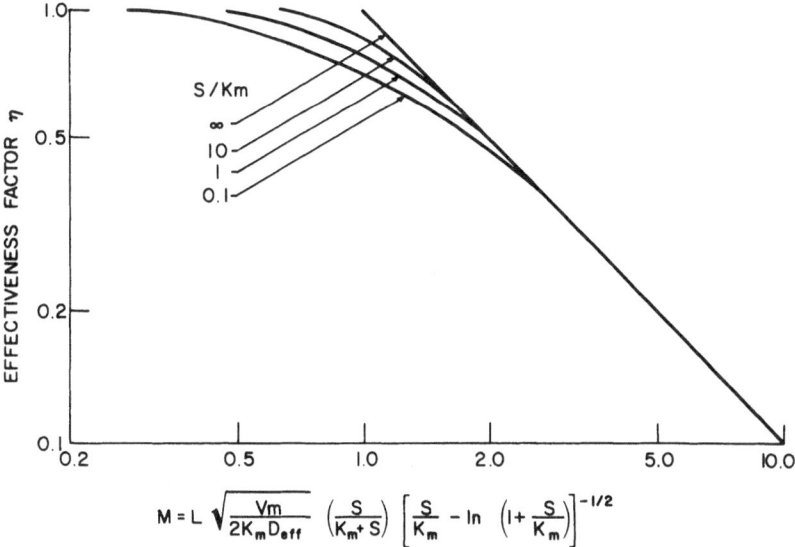

$$M = L \sqrt{\frac{Vm}{2K_m D_{eff}}} \left(\frac{S}{K_m + S}\right) \left[\frac{S}{K_m} - \ln\left(1 + \frac{S}{K_m}\right)\right]^{-1/2}$$

Fig. 4. Effectiveness factor for Michaelis-Menten kinetics

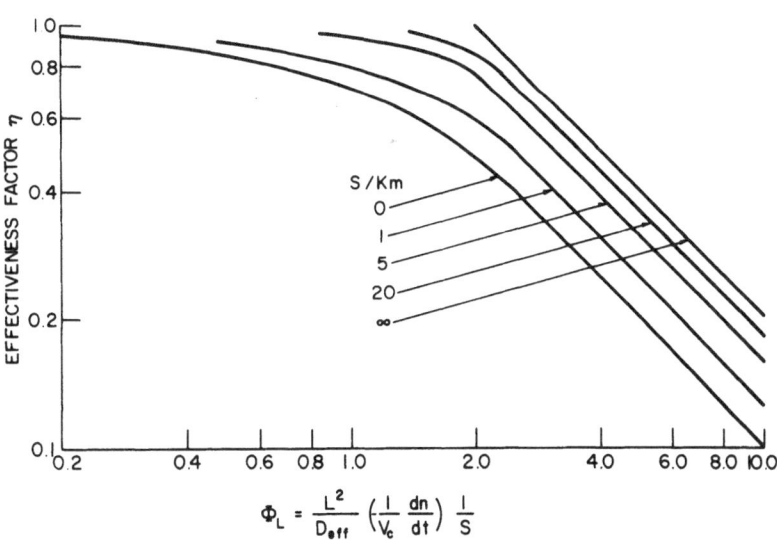

$$\Phi_L = \frac{L^2}{D_{eff}} \left(\frac{1}{V_c} \frac{dn}{dt}\right) \frac{1}{S}$$

Fig. 5. Effectiveness factor for Michaelis-Menten kinetics, flat plate geometry

For either type of modulus, perhaps the greatest problem is to determine or estimate the effective diffusivity. The tortuosity, which must be found experimentally or esti- mated from data on similar materials, is usually the most difficult factor to determine. If the diameter of the diffusing substrate molecules is a significant fraction of the pore diameter, the effective diffusivity will be decreased. Pitcher [54] and Satterfield *et al.* [55] proposed a correlation that can be used to estimate the resulting effective diffu- sivity.

Recent advances in theoretical treatments of diffusion have come in combining analysis of internal and external mass transfer effects. Hamilton *et al.* [56] provided a review of appropriate literature as well as their own work on combined internal and external diffusion effects. Frouws *et al.* [57] have used a graphical method for calculating con- version under combined intra- and extra-particle diffusion limitations.

Kobayashi and Laidler [58] devised a theoretical treatment of flow systems inside tubes lined with a porous layer containing immobilized enzyme. The group headed by Laidler has tested this theory experimentally on several occasions [59–61]. The theory appears to hold at least under limiting conditions.

Specialized treatment of hollow fiber reactor performance has also been reported [62]. Since in many cases the theoretical analysis of internal diffusion effects is either extreme- ly complicated or depends on parameters that are difficult to measure accurately, it is important to be aware of the practical methods of estimating these diffusion effects. One method is to decrease particle size or membrane thickness until no additional increase in observed reaction rate results. Marsh *et al.* [63] used this approach in deter- mining effectiveness factors for glucoamylase immobilized on porous glass. They com- pared reaction rates using immobilized enzyme particle of various sizes ranging down to finely crushed powder.

Another practical method of estimating pore diffusion effects is to compare the varia- tion of reaction rate or activity with temperature for the immoblized enzyme versus the soluble enzyme. If an Arrhenius plot (log of reaction rate versus reciprocal absolute temperature) for the immobilized enzyme gives a straight line with a slope equal to that of a similar plot for the soluble enzyme, it usually implies that no pore diffusion limitations exist. Internal diffusion limitations will normally result in a decrease in slope at higher temperatures. Wheeler [64] showed that as the effectiveness factor falls below unity, the apparent activation energy also decreases, approaching the arithmetic mean of activation energies for the chemical reaction and diffusion. Since the activation energy of diffusion is almost always small, the observed activation energy will approach half the intrinsic value as diffusion becomes the limiting factor. Havewala and Pitcher [65] and more recently, Buchholz and Ruth [66] have reported observing such behavior.

Another much less reliable indication of internal diffusion limitations is the observation of substantial reductions in enzyme activity upon immobilization. Since there are other causes of reduced activity, caution must be used in interpreting this type of observa- tion. On the other hand, agreement between soluble and immobilized enzyme specific activities at their respective pH optimal does indicate that diffusion resistance is low, as well as that the coupling method is efficient.

Initial rate assays are affected by pore diffusion limitations also. For example, the slope of Lineweaver-Burk plots becomes steeper as diffusion resistance increases. Bunting and Laidler [59] give such an example with varying amounts of β-galactosidase trapped in polyacrylamide gel.

4.2 External Mass Transfer

Several methods can be used to estimate the effects of external mass transfer limitations on apparent reaction rates.

The rate of mass transfer to a surface from the bulk solution can be written as

$$v = K_m a_m (S_b - S_s), \tag{10}$$

where K_m is the mass transfer coefficient (with units cm/s), a_m is the surface area per unit volume (cm^{-1}), S_b is the bulk phase substrate concentration, and S_s is the surface substrate concentration. This mass transfer rate, under steady-state conditions, must equal the reaction rate at the particle surface or the apparent rate within the particle. The mass transfer coefficient can be estimated from a number of proposed correlations. One of the most broadly based and recent correlations is that proposed by Wilson and Geankoplis [67]:

For $0.0016 < N_{Re} < 55 <$ and $0.35 < \epsilon < 0.75$

$$J = (1.09/\epsilon)N_{Re}^{-2/3}. \tag{11}$$

For $55 < N_{Re} < 1500$

$$J = (0.250/\epsilon)N_{Re}^{-0.31}, \tag{12}$$

where $N_{Re} = d_p G/\mu$, $J = K_m \rho/G)N_{Sc}^{2/3}$, $N_{Sc} = \mu/D_0$, μ is the liquid viscosity, G is the mass velocity per unit superficial bed cross-section, ρ is the liquid density, D_0 is the substrate diffusivity, ϵ is the void fraction, and d_p is the particle diameter.

For the lower Reynold's number range, usually encountered with immobilized enzyme systems, these correlations would suggest that the mass transfer coefficient should be proportional to flow rate to the one-third power. Rovito and Kittrell [47] found the best data fit with flow rate to the 0.5 power. Raman et al. [68] found their reaction rates in tubular reactors proportional to the one-third power of flow rate.

Satterfield [38, 69] has suggested another method for estimating the effect of film diffusion in a fixed-bed reactor. From single-point reactor performance data, the height of bed required for the necessary mass transfer to occur assuming the film diffusion to be the rate controlling step. This bed height (z) can be estimated from the correlation

$$z = \frac{\epsilon N_{Re}^{2/3} N_{Sc}^{2/3}}{1.09 \, a_v} \ln \frac{Y_1}{Y_2}, \tag{13}$$

where a_v is the ratio of particle external surface area to reactor volume, Y_1 is the mole fraction substrate in the feed, Y_2 is the mole fraction substrate in the product, and ϵ

is the void fraction. Havewala and Pitcher [23] give an example of such a calculation for an immobilized glucose isomerase fixed-bed reactor of 12.8 cm bed height operated at 100 ml/h feed rate (50% glucose solution) yielding 45% conversion. The bed height, calculated from Eq. (13) assuming film diffusion control, was 0.23 cm. This height is much less than the actual bed height indicating that the film diffusion resistance is insignificant.

Another practical method of testing for external mass transfer effects in fixed-bed reactors is to change the bed depth while keeping the residence time the same. If film diffusion effects are significant, the change in linear velocity will cause a change in conversion by the reactor even with the same residence time. Care must be exercised to examine a sufficient range in linear velocity to result in an observable effect if present. Satterfield [38] discussed the problem of mass transfer in slurry reactors (fluidized-bed or slurry reactors), citing the correlation from Brian and Hales [70].

$$\left(\frac{k_m d_p}{D}\right)^2 = 4.0 + 1.21 \ N_{Pe}^{2/3}, \tag{14}$$

where $N_{Pe} = d_p u/D$ and u is the fluid velocity. He also gave the correlation

$$k_m' \ (N_{Sc})^{2/3} = 0.38 \left(\frac{g\mu\Delta\rho}{\rho^2}\right)^{1/3} \tag{15}$$

where k_m' is the value of k_m for a sphere setting at its terminal velocity, $\Delta\rho$ is the difference between particle and fluid (ρ) densities, and g is gravitational acceleration. In stirred-tank reactors, particle entrainment in the fluid can limit the mass transfer rate even if agitation is increased. O'Neill [71] has also discussed the problem of mass transfer in stirred batch reactors. Horvath and Solomon [72] reported experimental data and developed a theoretical treatment of bulk diffusion controlled reactions using reactors consisting of tubes with enzymes immobilized on the inner wall. Similar studies have been described by Kobayashi and Laidler [58] and Bunting and Laidler [59]. In the area of packed-bed reactors, Kobayashi and Moo-Young [53, 73] have carried various theoretical and experimental studies including non-idealities in reactor performance.

A curious and not totally explained phenomenon, apparently related to contacting or mass transfer limitations, has been noted in fluidized-bed systems. In the hydrolysis of lactose [17] and the hydrolysis of cellulose [74] the extent of reaction for a given contact or residence time varies with flow rate. Low conversions were observed at both low and high linear velocities with a maximum at an intermediate flow rate, perhaps related to flow pattern changes.

4.3 Transient Behavior of Immobilized Enzyme Reactors

The assumption of steady state operation is applicable to most industrial reactor systems. There are, however, a few instances in which analysis of transient behavior could be valuable. In certain analytical applications of immobilized enzymes where small packed columns are used, response time and recovery time between samples are dependent upon transient behavior. For industrial reactors, cleaning cycles or other regular start-up and shut-down procedures could result in transient behavior significant to overall operating economics. Temporary perturbations in flow rate, temperature, pH, or feed concentration may also have important effects on reactor operation. Gellf et al. [75] solved numerically the case of transient behavior in an ideal plug-flow reactor (with no mass transfer limitations) subject to Michaelis-Menten kinetics. Experimental data from a column of immoblized α-chymotrypsin was compared with the numerical solutions for cases of varying substrate concentration. Ryu et al. [76] analyzed the transient behavior of a continuous stirred-tank reactor. They then extended this analysis to the plug flow case by considering a multi-stage continuous stirred tank reactor system.

Shyam et al. [51] included mass transfer limitations and transient response in a dynamic reactor model. Vieth et al. [3] have also reviewed the transient analysis of immobilized enzyme reactors.

4.4 Electrostatic Effects

Electrostatic effects have usually been mentioned only when a qualitative explanation for some phenomenon such as pH optima shifts was required. Early attempts to treat this effect quantitatively included the approach of Hornby et al. [77], who lumped diffusive and electrostatic effects together by using a modified Michaelis constant. Shuler et al. [78] quantitatively treated the effects of electrostatic fields and diffusion on the rates of reactions promoted by immobilized enzymes. They based their approach on a distribution of potential for the electrical double layer valid for small surface potential. Hamilton et al. [79] then extended this treatment to the higher surface potential case by employing the complete Guoy-Chapman solution. The results of Shuler et al. remain sufficiently accurate for most practical cases. Hamilton et al. [79] discuss the implications of the theoretical predictions by showing examples of the curvature of Lineweaver-Burk plots resulting from significant diffusional and electrical effects. They also make the interesting point that for substrate and surfaces of opposite charge, the effectiveness factor can exceed unity. Kobayashi and Laidler [80] have extended their work to the area of electrostratic and diffusive effects in membrane systems. Karube et al. [81] recently have reported affecting immobilized lipase activity by use of electric fields.

4.5 Backmixing

A common source of inefficiency in reactors that attempt to achieve plug flow conditions is backmixing. The significance of this backmixing, which is encountered in all real reactors, can be estimated from a calculated dispersion number as described by Levenspiel [82]. The dispersion number is defined as D_c/uL_c where D_c is the dispersion coefficient, u is the interstitial velocity and L_c is the bed depth. In the literature reference is sometimes made to the Bodenstein number or uL_c/D_c, the reciprocal of the dispersion number.

The dispersion number can be written in terms of the Peclet number ($N_{Pe} = u d_p/D_c$):

$$\frac{D_c}{uL_c} = \frac{d_p}{L_c N_{Pe}}. \tag{16}$$

Chung and Wen [83] developed a correlation based on extensive experimental data, between the Peclet number and the Reynolds number (N_{Re}) for fixed and fluidized beds.

$$\frac{N_{Pe}\epsilon}{Z} = 0.20 + 0.011 \, N_{Re}^{0.48}, \tag{17}$$

where $Z = 1$ for a fixed bed, $Z = (N_{Re})_{mf}/N_{Re}$ for fluidized beds, and $(N_{Re})_{mf} =$ the minimum Reynolds number for fluidization. This minimum fluidization Reynolds number can either be measured experimentally or estimated from the following equation:

$$(N_{Re})_{mf} = [(33.7^2 + 0.0408 \, N_{Ga})^{1/2} - 33.7] \tag{18}$$

where the Galileo number $(N_{Ga}) = [d_p^3 (\rho_s - \rho)g]/\mu^2$, ρ is the fluid density, ρ_s is the particle density, g is the gravitational constant, and μ is the fluid viscosity. For Reynolds numbers less than 10 in fixed beds, the Peclet number is approximately equal to 0.5. This result can be obtained from Eq. (17) by neglecting the Reynolds number term (since it will be very small) and by assuming a void fraction of 0.4.

At high substrate concentrations ($S \gg K_m$), reactions following Michaelis-Menten kinetics become essentially zero-order and conversions are unaffected by backmixing. The reaction becomes first-order, a case considered by Levenspiel, at low substrate concentrations. At intermediate substrate concentrations for Michaelis-Menten kinetics, Kobayashi and Moo-Young [73] give generalized plots comparing reactor sizes for actual versus plug flow conditions in the same manner as developed by Levenspiel. Tracer study data for a packed bed has been reported by Levenspiel [82]. Unfortunately, channeling and other irregularities in liquid-carrier contacting, not particularly amenable to accurate analysis, appear to be of more concern than backmixing [85].

5 Heat Transfer

In cases where immobilized enzyme performance (activity and stability) is sensitive to temperature, close temperature control in the reactor is important to insure operation at optimal conditions. Generally, relatively good heat transfer and temperature control can be achieved in fluidized-bed and stirred-tank reactors. Fixed-bed reactors present a more difficult problem. Havewala and Pitcher [23] estimated the heat transfer rate in a bed packed with immobilized glucose isomerase. They concluded that, for a typical diameter of 15 cm or more, an adiabatic temperature change resulting from the heat of reaction would occur, at least in the center of the column, in spite of any external insulation or even reasonable levels of heating. More recently, Marsh and Tsao [84] experimentally observed temperatures in a packed bed reactor containing glucoamylase immobilized on porous glass. They used a 4.2 cm inside diameter Plexiglass column, 1.8 m tall, surrounded with 3.1 cm of insulation. At a flow rate of 44.5 ml/minute, the feed temperature dropped from 55 to 51 °C using water and to 50 °C using 27% maltose feed. A similar one degree difference due to heat of reaction was observed for a 27% solution of 15 DE starch. It is worth noting that the heat of reaction for enzyme promoted reactions is frequently low in absolute magnitude.

6 Temperature Effects

Both enzyme promoted reaction rates and enzyme activity loss rates vary with temperature. Generally, their temperature dependence can be described by the Arrhenius equation. Activation energies are usually in the 5 to 20 kcal/g mole range while deactivation energies of 10 to more than 100 kcal/g mole can be observed. As an example, Reilly [86] reported an activation energy of 6.9 kcal/g for dextransucrase immobilized on alkylamine porous silica. From the immobilized enzyme stability data also reported, a deactivation energy of about 63 kcal/g mole can be estimated.

7 Immobilized Enzyme Activity Loss

Enzyme activity loss during operation is one of the most important factors in determining system performance and economics. Activity loss can result when the enzyme leaches out of the system, the enzyme itself becomes inactive, or the immobilized enzyme becomes coated or blocked from contact with the substrate.

The most commonly observed activity loss pattern is exponential decay where a plot of the logarithm of activity versus time is a straight line. Plots of this type are also occasionally downward or upward concave. One cause of the latter case is initial enzyme washout from the carrier over a limited period of time. The former case is sometimes caused by cumulative poisoning. This downward concave decay curve has been reported by Lee [18] and by Beck and Rose [87].

Carrier attrition has been cited as a cause of activity loss [88]. There has also been speculation that shearing could remove enzyme from carriers. However, Weetall et al. [89] calculated that the disruption of covalent bonds by shearing in fixed-bed operation is practically impossible.

Several investigators [90–92] have attempted to improve glucose oxidase and catalase stability by controlling hydrogen peroxide levels. O'Neill [93] has also used lactate dehydrogenase as an example in discussing the effect of substrate protection and reactor type on immobilized enzyme life.

It is important to note that immobilized enzymes whose apparent activity is limited by diffusion will exhibit slower apparent loss of activity than if no diffusion limitations were present. As the enzyme activity is lost, the effectiveness factor increases, thus slowing the apparent activity loss rate. Ollis [94] has discussed this phenomenon in greater detail.

8 Operating Strategy

This author [1, 2] has previously discussed briefly immobilized enzyme reactor operating strategy. Potentially important to immoblized enzyme system economics, this area has received little attention.

Usually it is desirable to maximize the total amount of feed processed per unit of enzyme, reactor volume, or some such variable with the ultimate goal of minimizing total processing cost.

The total production P_t of a reactor during a period of time t_p can be related to the feed rate F by the equation

$$P_t = \int_0^{t_p} F \, dt. \tag{19}$$

For the most common case, exponential activity decay and operation at constant conversion,

$$P_t = \int_0^{t_p} F_i \exp\left[-(\ln 2)t/t_{1/2}\right] dt, \tag{20}$$

where F_i is the initial feed rate and $t^{1/2}$ is enzyme half-life. After integration is performed, this equation becomes

$$P_t = \frac{F_i t_{1/2}}{\ln 2} (1 - \exp\left[-(\ln 2) \, t_p/t_{1/2}\right]). \tag{21}$$

Operating at constant conversion implies decreasing flow rates or production rates to maintain conversion as enzyme activity decreases. This change in production rate during the life of the immobilized enzyme is frequently unacceptable for a commercial process. The most common solution to this problem is to utilize a multiple-reactor system. These reactors with staggered start-up or reloading times can be operated in series or parallel. The variation in production rate can be held to any specified level by using a sufficient number of reactors. The number of reactors required to maintain the production rate within given tolerance levels is a function of the number of half-lives for which the reactor is operated before the immobilized enzyme is replaced [23].

$$R_p = e^{\frac{-H \ln 2}{N}}, \tag{22}$$

where N is the number of reactors, H is the number of half-lives for which the immobilized enzyme is used, and R_p is the ratio of the maximum to the minimum production rate. An example is shown in Fig. 6 where $R_p = 0.82$, N = 7, and H = 2. This strategy is the optimal one if the percentage decrease in half-line with temperature is greater than the percentage increase in activity. The minimum allowable temperature is then selected as the operating temperature.

Another strategy of column operation is to raise the operating temperature to compensate for activity losses, thus maintaining the original production rate and conversion level.

Pitcher [95] discussed a case where other constraints are placed on strategy selection. These limits included upper and lower temperature limits, frequently encountered because of microbial contamination problems and product quality considerations, and a fixed operating time. The optimal operating strategy was determined for an actual immobilized lactase system assuming 40 and 50 °C temperature limits and a 300-day operating period. The optimal temperature policy was found to involve

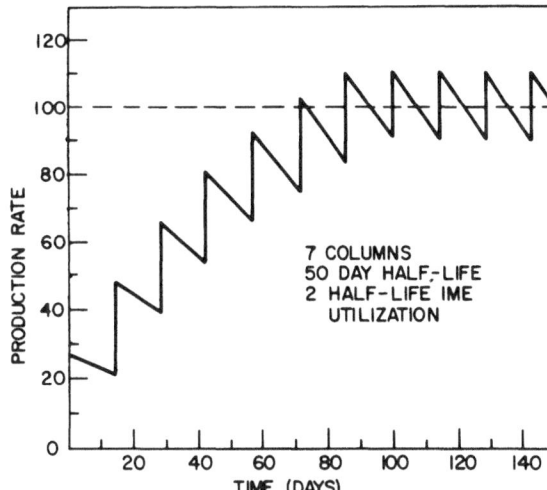

Fig. 6. Production rate for
multiple reactor system

three distinct phases. The reactor must initially be operated at the minimum temperature, in this case 40 °C. Then constant conversion must be maintained with the
temperature being increased until the maximum temperature, 50 °C, is reached. The
final phase involves isothermal, 50 °C, operation. The flow rate (production rate)
as a function of time is shown in Fig. 7. An approximation of this operating strategy
that achieved productivity of only 0.1% less is also shown. For actual system operation
the large drop in production rate during the final phase may be unattractive. A
strategy involving only 40 °C and constant conversion operation, for 168 and 132 days
respectively, may be more practical and yield only 1% less total productivity. A more
theoretical treatment of this type of reactor operating strategy problem is given by

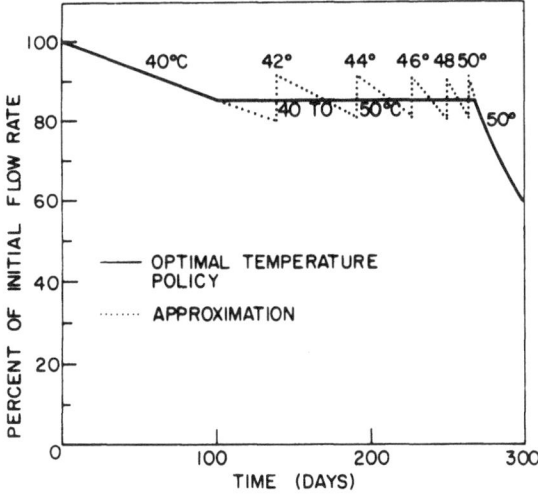

Fig. 7. Optimal temperature policy

Crowe [96]. Haas *et al.* [97] have reported using the calculus of variation to find the optimal temperature policy for a first-order reversible batch reaction.

9 Multienzyme Systems

Although numerous studies have been made of multi-enzyme systems [45, 98–105], where as many as four enzymes [106] have been acting simultaneously, little attention has been paid to optimization. Goldman and Katchalski [107] performed a theoretical kinetic analysis of a two-enzyme system. Ford [108] analyzed the oxidation of ethanol to acetic acid by alcohol and aldehyde dehydrogenases in a hollow fiber reactor. He determined the optimal enzyme ratio for the overall conversion of ethanol to acetic acid. Kent and Emery [109] combined immobilized glucoamylase and glucose isomerase in varying proportions and studied the conversion of maltose to fructose at various pH levels. Reilly [110] also used the same two-enzyme system for conversion of dextrin to fructose. He also discussed the problem of optimizing catalyst profiles for two step reactions. Ho and Kostin [111] have derived expressions for the concentration of an immobilized multi-enzyme system in which consecutive reactions and diffusion occur.

10 System Cost Estimates

Preliminary cost estimates made to evaluate the feasibility of an immobilized enzyme system can be broken down into several major components. The first is the cost of the immobilized enzyme itself. This cost, in turn, consists of carrier, enzyme and immobilization costs including labor, capital and materials. From the immobilized enzyme cost and the performance data relating total production to the amount of immobilized enzyme used, the processing cost attributable to the immobilized enzyme can be calculated per unit of product.

A second cost component is labor for operating and maintaining the process. Supervision and other overheads may bring this cost to 2 or 3 times the hourly wage rate or more. The third cost component, equipment cost, appears as depreciation. Taxes, insurance, and maintenance are also frequently estimated as a percentage of capital. This total charge may exceed 20% annually.

To obtain a product cost, raw material costs must also be added.

Examples of preliminary cost estimates for immobilized enzyme systems are rarely found in the literature. Pitcher and Weetall [112] and Pitcher *et al.* [113] have given such estimates for glucoamylase and lactase immobilized enzyme systems, respectively. Other estimates appearing in the literature have generally been even more superficial than these preliminary estimates.

References on cost estimating are available [114, 115]. As long as the proper information concerning immobilized enzyme performance is used and realistic allowance is made for immobilized enzyme losses in preparation and handling, the preliminary cost estimation procedure should not be complicated.

11 General Design Considerations

This author has previously [1, 2] reviewed some of the general considerations influencing immobilized enzyme reactor system design. Although these factors should be well-known by now, they are probably worth reviewing.

The immobilized enzyme characteristics and reactor design parameters already considered individually are not really independent and affect each other as well as the ultimate efficiency or cost of the whole system.

For example, the reaction rate or rate of product formation depends not only upon enzyme kinetics and substrate concentration, but also upon carrier geometry (mass transfer limitations), operating temperature, pH, and contaminants or activators in the feed.

The amount of enzyme immobilized per unit volume or weight of carrier affects the processing cost as a function of carrier cost. For low cost carriers, high enzyme loading is less important. For low cost enzymes high loading is necessary to reduce the impact of carrier cost. Enzyme loading also affects reactor size and equipment costs. Enzyme loading itself depends upon the immobilization technique conditions during attachment, and carrier surface area, pore size, and composition.

Coupling efficiency, the percentage of enzyme activity not recovered from the enzyme solution after immobilization that is observed as immobilized enzyme activity, also depends upon the same variables affecting loading. In general, however, the more enzyme immobilized on a given amount of carrier, the lower the coupling efficiency. Eaton et al. [116] gave an example of immobilized lactase. Yamane [117] has also discussed this subject.

Pitcher [95] has reported apparent half-lives for immobilized lactase ranging from 2 days to over 100 days, depending on feed composition. During development studies it is essential to duplicate as closely as possible the feed projected for commercial use. During long-term continuous operation, microbial growth or trace contaminants such as heavy metals can severely decrease enzyme activity.

The prevention or control of microbial contamination, of grave concern commercially, has not received much attention. Barndt et al. [118] evaluated the effect of various sanitizing agents including iodophor, hydrogen peroxide, an acid cleaner, and a quaternary amine on a lactase-collagen complex. Pitcher et al. [113] reported that periodic backflushing with dilute acetic acid successfully controlled microbial contamination in immobilized lactase columns. Several instances [119, 120] of the use of immobilized enzymes themselves as bacteriacidal agents have been reported.

Pressure drop and plugging problems, encountered primarily in fixed-bed reactors, depend upon a number of factors including feed rate, bed height, feed composition and temperature (affecting viscosity), carrier particle size, and bed packing characteristics. Pressure drop is extremely sensitive to the bed void fraction [1, 2]. A trade-off between pore diffusion problems and pressure drop must sometimes be made.

Use of immobilized enzymes may necessitate or eliminate additional steps. For example, product discoloration or other time-dependent side reactions can be avoided by the short residence times associated with at least some immobilized enzyme reactor systems.

On the other hand, electrolyte activators introduced into the feed to enhance enzyme activity or stability must sometimes be removed by ion exchange or other treatment.

12 Future Trends

As stated in the introduction of this article, immobilized enzyme reactor design and operation is not a rapidly changing field. Most of the basic concepts have probably been established and remarkable gains in operating efficiency seen unlikely in most cases. New methods of obtaining higher purity feed streams or avoiding microbial contamination may yet bring additional system applicability. However, reactor engineering should probably be viewed primarily as a tool to enable proper evaluation and utilization of new immobilized enzyme systems.

13 Table of Symbols

a_m	surface area per unit volume
a_v	ratio of particle surface area to reactor volume
d_p	particle diameter
D	substrate diffusivity
D_{eff}	effective diffusivity $= D\Theta/\tau$
D_0	bulk diffusivity
E	enzyme activity
E_t	total enzyme activity
F	flow rate
F_i	initial feed rate
g	acceleration due to gravity
G	mass velocity per unit superficial bed cross section
H	number of half-lives utilization of IME
J	dimensionless group
k	turnover number
k_m	mass transfer coefficient
k'_m	value of k_m for a sphere setting at its terminal velocity
k_v	reaction velocity constant
K	equilibrium constant
K_i	product inhibition constant
K_m	Michaelis constant
K'_m	substrate inhibition constant
K_p	constant (Michaelis type for reverse reaction)
L	flat plate thickness
L_c	bed height
m	order of reaction
M	general modulus
N	number of reactors
N_{Ga}	Galileo number $= d_p^3 \rho(\rho_s - \rho)g/u^2$
N_{Pe}	Peclet number $= d_p u/D$
N_{Re}	Reynolds number $= dpG/\mu$
$(N_{Re})_{mf}$	minimum fluidization Reynolds number
N_{Sc}	Schmidt number $= \mu/\rho D$

P	product concentration
P_t	total production
r	radial distance
R_p	ratio of low to high production rate
S	substrate concentration
S_b	bulk substrate concentration
S_0	initial substrate concentration
S_s	substrate concentration at catalyst surface
S_t	total substrate concentration if all product coverted to substrate
t	reaction time
$t_{1/2}$	enzyme half life
t_p	total period of time of reactor operation
u	fluid velocity
v	reaction velocity
v_i	intrinsic reaction rate
V_m	kE (maximum reaction velocity)
V_s	substrate volume
W	weight of immobilized enzyme
X	$(S_0 - S)/S_0$ = fractional conversion
X_e	X_t at equilibrium
X_i	$\dfrac{S_t - S_0}{S_t}$
X_t	$\dfrac{S_t - S}{S_t}$
Y_1	mole fraction substrate in feed
Y_2	mole fraction substrate in product
z	column height, assuming film diffusion to the rate controlling step
Z	$(N_{Re})_{mf}/R_{Re}$
ϵ	void fraction
η	effectiveness factor
OL	modulus
μ	fluid viscosity
ρ	fluid density
ρ_s	solid density
Θ	particle internal porosity
τ	tortuosity (ratio of actual diffusion path length to straight line distance)

14 References

1. Pitcher, W. H. Jr.: Catal. Rev. – Sci. Eng. **12(1)**, 37 (1975)
2. Pitcher, W. H. Jr.: In Immobilized Enzymes for Industrial Reactors. R. A. Messing (Ed.), p. 151. New York: Academic Press, Inc. 1975
3. Vieth, W. R., Venkatasubramanian, K., Constantinides, A., Davidson, B.: In: Applied Biochemistry and Bioengineering, p. 221. New York: Academic Press, Inc. 1976
4. O'Neill, S. P., Wykes, J. R., Dunnill, P., Lilly, M. D.: Biotechnol. Bioeng. **13**, 319 (1971)
5. Coughlin, R. W., Aizawa, M., Charles, M.: Biotechnol. Bioeng. **18**, 199 (1976)
6. Robinson, P. J., Dunnill, P., Lilly, M. D.: Biotechnol. Bioeng. **15**, 603 (1973)
7. Havewala, N. B., Weetall, H. H.: U. S. Pat. 3 767 535 (1973)
8. Dinelli, D.: Process Biochem. 9 (August 1972)
9. Morisi, F., Pastore, M., Viglia, A.: J. Dairy Sci. **56**, 1123 (1973)
10. Marconi, W., Gulinelli, S., Morisi, F.: Biotechnol. Bioeng. **16**, 501 (1974)
11. Closset, G. P., Shah, Y. T., Cobb, J. T.: Biotechnol. Bioeng. **15**, 441 (1973)

12. Closset, G. P., Cobb, J. T., Shah, Y. T.: Biotechnol. Bioeng. **16**, 345 (1974)
13. Emery, A. H.: In: Enzyme Engineering, Vol. 2, E. K. Pye, L. B. Wingard, Jr. (Eds.), P. 269. New York: Plenum Press 1974
14. Venkatasubramanian, K., Vieth, W. R.: Biotechnol. Bioeng. **15**, 583 (1973)
15. Horvath, C., Solomon, B. A., Engasser, J. M.: Ind. Eng. Chem. Fundam. **12**, 431 (1973)
16. Coughlin, R. W.: U. S. Pat. 3928143 (1975)
17. Charles, M., Coughlin, R. W., Allen, B. R., Paruchuri, E. K., Hasselberger, F. X.: Increasing Economic Value of Whey Waste waters Using Immobilized Lactase. Paper 17b, presented at AIChE 66th Annual Meeting, Philadelphia (1973)
18. Lee, Y. Y., Wun, K., Tsao, G. T.: Kinetics and Mass Transfer Characteristics of Glucose Isomerase Immobilized on Porous Glass. Paper 11a, presented at the AIChE 77th National Meeting, Pittsburgh (1974)
19. Emery, A. N., Revel. Chion, L.: Hydrodynamic Effects in the Design of Fluidized Bed Immobilized Enzyme Reactors. Paper 11b, presented at the AIChE 77th National Meeting, Pittsburgh (1974)
20. Shumate, S. E.: Multi-Stage Tapered Fluidized Beds as Bioreactors. Presented at Enzyme Engineering Conference, Portland, Oregon (August 1975)
21. Gellf, G., Boudrant, J.: Biochim. Biophys. Acta **334**, 467 (1974)
22. Lilly, M. D.: Comments at Enzyme Engineering Conference, Bad Neuenahr, Germany (1977)
23. Havewala, N. B., Pitcher, W. H.: In: Enzyme Engineering, Vol. 2, E. K. Pye, L. B., Wingard, Jr. (Eds.), p. 315. New York: Plenum Press 1974
24. Dixon, M., Webb, E. C.: Enzymes. New York: Academic Press, Inc. 1964
25. Plowman, K. M.: Enzyme Kinetics. New York: McGraw-Hill 1972
26. Reiner, J. M.: Behavior of Enzyme Systems. New York: Van Nostrand Reinhold 1969
27. Westley, J.: Enzymic Catalysis. New York: Harper & Row 1969
28. Atkins, G. L., Nimmo, I. A.: Biochem. J. **135**, 779 (1973)
29. Bates, D. J., Fridean, C.: J. Biol. Chem. **248**, 7878 (1973)
30. Bates, D. J., Friedan, C.: J. Biol. Chem. **248**, 7885 (1973)
31. Eisenthal, R., Cornish-Bowden, A.: Biochem. J. **139**, 715 (1974)
32. Cornish-Bowden, A., Eisenthal, R.: Biochem. J. **139**, 721 (1974)
33. Yun, S.–L., Suelter, C. H.: Biochim. Biophys. Acta **480**, 1 (1977)
34. Cornish-Bowden, A.: Biochem. J. **149**, 305 (1975)
35. Storer, A. C., Darlison, M. G., Cornish-Bowden, A.: Biochem. J. **151**, 361 (1975)
36. Fjellstedt, T. A., Schlesselman, J. J.: Anal. Biochem. **80**, 224 (1977)
37. Duggleby, R. G., Morrison, J. F.: Biochim. Biophys. Acta **481**, 297 (1977)
38. Satterfield, C. N.: Mass Transfer in Heterogeneous Catalysis. Cambridge, Massachusetts: MIT Press 1970
39. Blaedel, W. J., Kissel, T. R., Boguslaski, R. C.: Anal. Chem. **44**, 2030 (1972)
40. Fink, D. J., Na, T. Y., Schultz, J. S.: Biotechnol. Bioeng. **15**, 879 (1973)
41. Gondo, S., Sato, T., Kussuroki, K.: Chem. Eng. Sci. **28**, 1773 (1973)
42. Kasche, V., Lundquist, H., Bergman, R., Axen, R.: Biochem. Biophys. Res. Commun. **45**, 615 (1971)
43. Kobayashi, T., Laidler, K. J.: Biochim. Biophys. Acta **302**, 1 (1973)
44. Kobayashi, T., Moo-Young, M.: Biotechnol. Bioeng. **15**, 47 (1973)
45. Lawrence, R. L., Okay, V.: Biotechnol. Bioeng. **15**, 217 (1973)
46. Mogensen, A. D., Vieth, W. R.: Biotechnol. Bioeng. **15**, 467 (1973)
47. Rovito, B. J., Kittrell, J. R.: Biotechnol. Bioeng. **15**, 143 (1973)
48. Sundarum, P. V., Tweedale, A., Laidler, K. J.: Can. J. Chem. **48**, 1498 (1970)
49. Thomas, D., Broun, G., Selegny, E.: Biochimie **54**, 229 (1972)
50. Vieth, W. R., Mendiratta, A. K., Mogensen, A. K., Saini, R., Venkatasubramanian, K.: Chem. Eng. Sci. **28**, 1013 (1973)
51. Shyam, R., Davidson, B., Vieth, W. R.: Chem. Eng. Sci. **30**, 669 (1975)
52. Bischoff, K. B.: AIChE J. **11**, 351 (1965)

53. Moo-Young, M., Kobayashi, T.: Can. J. Chem. Eng. **50**, 162 (1972)
54. Pitcher, W. H. Jr.: Sc. D. Thesis, Mass. Inst. of Tech. (1972)
55. Satterfield, C. N., Colton, C. K., Pitcher, W. H. Jr.: AIChE J. **19**, 628 (1973)
56. Hamilton, B. K., Gardner, C. R., Colton, C. K.: AIChE J. **20**, 503 (1974)
57. Frouws, M. J., Vellenga, K., DeWilt, H. G. J.: Biotechnol. Bioeng. **18**, 53 (1976)
58. Kobayashi, T., Laidler, K. J.: Biotechnol. Bioeng. **16**, 99 (1974)
59. Bunting, P. S., Laidler, K. J.: Biotechnol. Bioeng. **16**, 119 (1974)
60. Ngo, T. T., Laidler, K. J.: Biochim. Biophys. Acta **377**, 317 (1975)
61. Narinesingh, D., Ngo, T. T., Laidler, K. J.: Can. J. Biochem. **53**, 1061 (1975)
62. Waterland, L. R., Michaels, A. S., Robertson, C. R.: AIChE J. **20**, 50 (1974)
63. Marsh, D. R., Lee, Y. Y., Tsao, G. T.: Biotechnol. Bioeng. **15**, 483 (1973)
64. Wheeler, A.: Advan. Catal. **3**, 249 (1951)
65. Havewala, N. B., Pitcher, W. H. Jr.: Unpublished report (1973)
66. Buchholz, K., Ruth, W.: Biotechnol. Bioeng. **18**, 95 (1976)
67. Wilson, E. J., Geankoplis, C. J.: Ind. Eng. Chem. Fundam. **5**, 9 (1966)
68. Raman, S. V., Horbett, T. A., Hoffman, A. S.: J. Mol. Cat. **2**, 275 (1977)
69. Satterfield, C. N.: Personal communication (1972)
70. Brian, P. L. T., Hales, H. B.: AIChE J. **15**, 419 (1969)
71. O'Neill, S. P.: Biotechnol. Bioeng. **14**, 675 (1972)
72. Horvath, C., Solomon, B. A.: Biotechnol. Bioeng. **14**, 885 (1972)
73. Kobayashi, T., Moo-Young, M.: Biotechnol. Bioeng. **13**, 893 (1971)
74. Karube, I., Tanaka, S., Shirai, T., Suzuki, S.: Biotechnol. Bioeng. **19**, 1183 (1977)
75. Gellf, G., Thomas, D., Broun, G.: Biotechnol. Bioeng. **16**, 315 (1974)
76. Ryu, D. Y., Bruno, C. F., Lee, B. K., Venkatasubramanian, K.: Proc. IV Int. Ferment. Symp., Ferment. Technol. Today 307 (1972)
77. Hornby, W. E., Lilly, M. D., Crook, E. M.: Biochem. J. **107**, 669 (1968)
78. Shuler, M. L., Aris, R., Tsuchiya, H. M.: J. Theor. Biol. **35**, 67 (1972)
79. Hamilton, B. K., Stockmeyer, L. J., Colton, C. K.: J. Theor. Biol. **41**, 547 (1973)
80. Kobayashi, T., Laidler, K. J.: Biotechnol. Bioeng. **16**, 77 (1974)
81. Karube, I., Yugeta, Y., Suzuki, S.: Biotechnol. Bioeng. **19**, 1493 (1977)
82. Levenspiel, O.: Chemical Reaction Engineering. New York: Wiley 1962
83. Chung, S. F., Wen, C. Y.: AIChE J. **14**. 857 (1968)
84. Marsh, D. R., Tsao, G. T.: Biotechnol. Bioeng. **18**, 349 (1976)
85. O'Neill, S. P., Dunnill, P., Lilly, M. D.: Biotechnol. Bioeng. **13**, 337 (1971)
86. Reilly, P. J.: Iowa State University Engineering Research Report, ERI-77176, 136 (1976)
87. Beck, S. R., Rose, H. F.: Ind. Eng. Chem. Prod. Res. Develop. **12**, 260 (1973)
88. Regan, D. L., Dunnill, P., Lilly, M. D.: Biotechnol. Bioeng. **16**, 333 (1974)
89. Weetall, H. H., Havewala, N. B., Garfinkel, H. M., Buehl, W. M., Baum, G.: Biotechnol. Bioeng. **14**, 201 (1972)
90. O'Neill, S. P.: Biotechnol. Bioeng. **14**, 201 (1972)
91. Cho, Y. K., Bailey, J. E.: Biotechnol. Bioeng. **19**, 157 (1977)
92. Cho. Y. K., Bailey, J. E.: Biotechnol. Bioeng. **19**, 769 (1977)
93. O'Neill, S. P.: Biotechnol. Bioeng. **14**, 473 (1972)
94. Ollis, D. F.: Biotechnol. Bioeng. **14**, 871 (1972)
95. Pitcher, W. H. Jr.: Immobilized Lactase for Whey Hydrolysis: Stability and Operating Strategy. Presented at Enzyme Engineering Conference, Bad Neuenahr, Germany (1977)
96. Crowe, C. M.: Can. J. Chem. Eng. **48**, 576 (1970)
97. Haas, W. R., Tavlarides, L. L., Wnek, J. J.: Optimal Temperature Policy for (First-Order) Reversible Reactions with Deactivation: Applied to Enzyme Reactors. Paper 39c, presented at AIChE 66th Annual Meeting, Philadelphia (1973)
98. Mosbach, K., Mattiasson, B.: Acta Chem. Scand. **24**, 2093 (1970)
99. Mattiason, B., Mosbach, K.: Biochim. Biophys. Acta **235**, 253 (1971)
100. Wilson, R. J. H., Kay, G., Lilly, M. D.: Biochem. J. **109**, 137 (1969)
101. Srere, P. A., Mattiasson, B., Mosbach, K.: Proc. Nat. Acad. Sci., U. S. **70**, 2534 (1973)

102. Broun, G., Thomas, G., Selegny, E.: J. Membrane Biol. 8, 313 (1972)
103. Martensson, K.: Biotechnol. Bioeng. 16, 567 (1974)
104. Newirth, T. L., Diegelman, M. A., Pye, E. K., Kallen, R. G.: Biotechnol. Bioeng. 16, 1089 (1974)
105. Mosbach, K., Mattiason, B.: In Methods in Enzymology, Vol. 44. K. Mosbach (Ed.), p. 453. New York: Academic Press 1976
106. Brown, H. D., Patel, A. B., Chattopadhyay, S. K.: J. Chromatogr., 35, 103 (1968)
107. Goldman, R., Katchalski, E.: J. Theor. Biol. 32, 243 (1971)
108. Ford, J. R.: Ph. D. Thesis, Tulane University, 1972
109. Kent, C. A., Emery, A. N.: Reactor Design for the Immobilized Enzyme System Amyloglucosidase-Glucose Isomerase. Paper 11e, presented at AIChE 77th Annual Meeting, Pittsburgh (1974)
110. Reilly, P. J.: Iowa State University Engineering Research Report, ERI-77176, 84 (1976)
111. Ho, S. P., Kostin, M. D.: J. Chem. Phys. 61, 918 (1974)
112. Pitcher, W. H. Jr., Weetall, H. H.: Enzyme Technol. Digest 4, 127 (1975)
113. Pitcher, W. H., Jr., Ford, J. R., Weetall, H. H.: In Methods in Enzymology, Vol. 44. K. Mosbach (Ed.), p. 792. New York: Academic Press 1976.
114. Popper, H. (Ed.): Modern Cost-Engineering Techniques. New York: McGraw-Hill 1970
115. Guthrie, K. M.: Process Plant Estimating Evaluation and Control. Solana Beach, Calif.: Craftsman Book Company of America 1974
116. Eaton, D. L., Ford, J. R., Pitcher, W. H. Jr.: The Use of Controlled-Pore Ceramic Bodies for Enzyme Immobilization. Paper 11d, presented at AIChE 77th Annual Meeting, Pittsburgh (1974)
117. Yamane, T.: Biotechnol. Bioeng. 19, 749 (1977)
118. Barndt, R. L., Leeder, J. G., Giacin, J. R., Kleyn, D. H.: J. Food Sci. 40, 291 (1975)
119. Karube, I., Suganuma, T., Suzuki, S.: Biotechnol. Bioeng. 19, 301 (1977)
120. Mattiason, B.: Biotechnol. Bioeng. 19, 777 (1977)

Biotechnology of Immobilized Multienzyme Systems

S. A. Barker and P. J. Somers
Department of Chemistry, University of Birmingham
Birmingham, B15 2TT, Great Britain

Contents

1 Introduction 27
2 Kinetic Analysis of Immobilized Multi-Enzyme Reaction Systems 28
3 Specific Sequential Enzyme Systems 34
4 Immobilized Whole Cells 45
5 Conclusion 47
6 References 47

1 Introduction

A wide diversity of immobilized multienzyme systems are now available if the widest interpretation of this concept is applied. The ingenuity of the research worker is boundless in the manner in which such enzymes may be presented. For example the following possibilities are available for cellular based systems:
(i) The whole cell either in the form of a microbial film or in the less restricted flocculated form where the cell is still capable of respiration and cell reproduction.
(ii) The whole cell in an encapsulation form that prevents cell reproduction.
(iii) Variations on (i) where an additional enzyme, or enzymes, is attached to the surface of the cell which while remaining viable is then incapable of reproduction.
(iv) Variations of (i) where due to the manner of flocculation the possibility of reproduction is denied.
(v) Cells of the above types where sequences of enzyme reaction are confined by the use of selective inhibitors or inactivated by heat treatment to leave selective enzymic action available.
(vi) Cells of the above types confined by the use of a single substrate or limited range of substrates.
(vii) Combinations of any of the above, including those that can participate in a symbiotic relationship.
While it is, in general, cheaper of an industrial process to adopt procedures of the type outlined in (i)–(vii) above for multienzyme systems, the main sphere of interest still lies in the popular concept of immobilized cell-free enzymes. This is in part due to the not unexpected lack of published information on commercial cellular based systems.

Immobilization of isolated cell-free enzymes provides an easier system for study and analysis and it is natural that these presentations have received most attention. Some of the different presentations that have been encountered are:

(viii) a physical mixture of individually immobilized enzymes, each within its own microenvironment,

(ix) a synthetic fabricated cell-like structure with one or more enzymes attached to the surface,

(x) a synthetic fabricated analogy of a biological membrane with one or more enzymes attached to each side of, or within, the membrane and so arranged that the products of the enzyme(s) on one side are substrates of the enzyme(s) on the other,

(xi) the replacement of the membrane in (x) with its analogous tubular device such as a hollow semipermeable fibre,

(xii) the analogous system to (viii) where the enzymes are immobilized within the same microenvironment,

(xiii) combinations of the above systems.

Since the fabrication of such systems obviously provides an additional cost factor, specific advantages must accrue from the use of such systems. The synthetic microenvironment may be employed to act in one or more of the following roles:

(a) to confer thermal stability on the enzyme,

(b) to confer a wider range of pH stability with respect to the bulk solution,

(c) to alter the apparent optimum pH of individual enzymes so that the sequence of enzymes will be compatible with respect to pH requirements and hence work together more effectively,

(d) to act as a separating device in order to displace enzyme equilibria,

(e) to act as a physical barrier between an enzyme that may act on the substrate of another enzyme in the sequence, and

(f) to prevent any particular non-enzyme additives such as coenzymes, activators or chelating agents from inhibiting other parts of the enzyme system.

In this review it is not intended to describe in detail methods of immobilization for enzymes since this is now an established technique. Obviously when choosing suitable methods for use with multiple enzymes additional constraints may be placed on the choice of method depending on the specific data being designed. Whilst the section of specific examples is intended to be reasonably comprehensive, in order to present a broad view of the advantages, disadvantages and objectives of such systems, the section on immobilized whole cells is only intended to illustrate the use of such systems.

2 Kinetic Analysis of Immobilized Multi-Enzyme Reaction Systems

The study of theoretical models for immobilized multi-enzyme reaction systems is of interest not only for an improved understanding of in vitro behaviour of enzymes, but also for its application to in vivo systems since many enzymes act in vivo whilst within membranes or attached to subcellular particles. An estimate of the commercial significance of an immobilized enzyme system requires a thorough knowledge of parameters effecting its behaviour under operational conditions. The kinetic behaviour of immobilized enzymes is different from that of the enzyme in solution because of the possible rate limitations imposed by diffusion. Also effects such as steric changes, or chemical modification of the active groups of the enzyme due to covalent bonding to the matrix, can cause modification of the intrinsic kinetics of the enzyme reaction. Investigations on the enzyme kinetics must be performed in order to ascertain whether the catalyst is subject to inhibition by reactants or products, or both. The manner in which the kinetics effects the efficiency of reactor systems can be described theoretically. However, it is important to decide the validity of the derived expressions for application to practical systems in view of possible deviations due to the heterogeneous nature of the catalytic reaction. The choice of a particular reactor system is in practice dependent on a number of factors, but ultimately a knowledge of the actual process kinetics affects the choice of reactor.

The use of a simple Michaelis-Menten relationship has been described [1–3] for studies of the process kinetics of immobilized enzymes. More complex kinetics are observed should inhibition of either product or reactant be important. For example in a continuously stirred tank reactor it is possible to obtain, in theory, more than one steady state when inhibition by reactant occurs. This may well limit the degree of conversion attainable. In general terms inhibition by substrate is more problematical in a plug flow reactor whereas inhibition by product is more serious in a continuously stirred tank reactor.

Recently detailed kinetic treatments, both theoretically [4] and practically [5, 6] based, have been presented for immobilized multi-enzyme systems wherein the enzymes catalyse consecutive reactions as a step towards the development of more efficient and complex enzyme reactors.

A number of approaches can be used to simulate the behaviour of multi-enzyme systems. The arrangement of enzymes in a sequence in a packed column or sequence of columns, whilst convenient does not provide the most efficient mode of presentation. Binding of two or more enzymes to the same matrix so that the substrate for the second enzyme is generated in situ as the first reaction step occurs is more efficient. The kinetic behaviour of the sequential arrangement is however simple, since only one enzyme is acting at any position in the column or in any one column of a sequential array of columns. An example of this system has been described by Brown et al. [7], using the sequence of enzymes shown in Fig. 1. Each enzyme was immobilized in a polyacrylamide gel but no overlap of enzymes was permitted.

D-glucose + ATP $\xrightarrow[\text{hexokinase}]{}$ D-glucose-6-phosphate + ADP

D-glucose-6-phosphate $\xrightarrow[\substack{\text{phosphoglucose}\\\text{isomerase}}]{}$ D-fructose-6-phosphate

D-fructose-6-phosphate + ATP $\xrightarrow[\substack{\text{phosphofructo}\\\text{kinase}}]{}$ D-fructose-1,6-diphosphate + ADP

D-fructose-1,6-diphosphate $\xrightarrow[\text{aldolase}]{}$ Dihydroxyacetone phosphate + D-glyceraldehyde-3-phosphate

Fig. 1. The four step sequential immobilized enzyme reactor described by Brown et al. [7]

The performance of consecutive reactions involving two or more enzymes in solution, in essentially a batch process, has been analysed with particular reference to coupled enzyme assays [8, 9]. The additional complication of diffusion, and other effects, with immobilized systems makes theoretical treatment more difficult. The simplest treatment is possible when the intrinsic reaction kinetics are first order. This assumption is generally valid when the Michaelis constants for the two reaction steps are greater than the bulk substrate concentrations. This requirement would be satisfied for low initial substrate concentration therefore, or towards the end of a reaction in a batch system. First order reactions can be treated analytically with simple geometries to give an expression of comparative efficiency which is essentially independent of substrate or product concentration in bulk solution. Goldman and Katchalski [10] analysed a simple set of consecutive reactions, $S \rightarrow P_1 \rightarrow P_2$, for a two enzyme system attached to an impermeable membrane. The kinetic behaviour of this system is affected by diffusion through the unstirred layer at the membrane-solution interface and the assumption is made that a quasi-stationary state is established at this interface. Thus the rate of flow of product formed by the first enzyme into the bulk of the solution equals the difference between the rate of its production and the rate of its consumption by the second enzyme. The possible effects of localised pH gradients and changes in dielectric constant were neglected and the two enzyme activities were assumed to be independent. For all of the hypothetical systems studied [impermeable membrane impregnated with one or two enzymes] the products of the enzyme 1 and 2 in the bulk of solution, P_b^1 and P_b^2 respectively, increased linearly with time during the initial part of the reaction. With enzymes of relatively high activity, the enzyme reactions were diffusion controlled; the activity of enzyme 1 being limited by the rate of diffusion of substrate from the bulk of solution, and the activity of enzyme 2 approached that exhibited by enzyme 1. In constrast, analysis of a homogeneous reactor revealed an initial lag period in the production of P_b^2 of duration controlled by catalytic and physical parameters. With the systems analysed the concentration of P_b^1 approached a limiting value with time, this state being reached more rapidly with increased enzyme activity with a consequent lower observed concentration of P_b^1. The most significant finding was the prediction that the rate of production of end product during the initial period of reaction was considerably greater with the immobilized enzyme system compared with the homogeneous solution reactor. The distinguishing feature is the existence of the unstirred (diffusion) layer at the membrane-solution interface. When such a layer is absent the immobilized enzyme system in contact with a solution is characterised by a kinetic pattern identical to that of a homogeneous system. As a result of the enzyme reaction, substrate and product concentration gradients are established across the unstirred layer. This leads to a lowered concentration of the first substrate compared with the bulk solution. Consequently P^1 and P^2 at the site of enzyme reaction are higher than in bulk solution (Fig. 2).

A study of a nucleoprotein coascervate containing hexokinase and polynucleotide diphosphohydrolase has shown qualitative agreement with these findings [11]. The product of hexokinase action ADP was utilised by polynucleotide diphosphohydrolase to give polyadenylic acid. The rate of synthesis of the polynucleotide by both enzymes, when concentrated in a droplet phase, was markedly higher than in homogeneous

1 Enzyme		2 Enzymes

Homogeneous Solution $P_b^1 = k_1' S_b t$

$$P_b^2 = \frac{k_1'}{k_2'} S_b [k_2' t + \exp(-k_2' t) - 1]$$

Attached to an Imperme-able Membrane $P_b^1 = \dfrac{Ak_1 S_b Dt}{(k_1 l + D)v}$

$$P_b^2 = \frac{Dk_1 S_b}{vk_2(k_1 l + D)}\left\{ Ak_2 t + V \exp\left[\frac{-Ak_2 Dt}{v(k_2 l + D)}\right] - v \right\}$$

where P^1, P^2 are product concentrations
     v   is solution volume
     l   thickness of 'unstirred' layer
     b   as subscript indicates bulk solution $(x > 1)$
     A   is surface area in contact with volume v
     k_1, k_2 are first order rate constants for enzymes 1 and 2
     k_1', k_2' are first order rate constants for enzyme 1 and 2 in solution
     D   is the diffusion coefficient

Fig. 2. Derived expression for product concentration with time in coupled enzyme systems for solution and immobilized situations [10]

solution or one in which the two enzymes were separated (hexokinase in solution and nucleotide diphosphohydrolase in the coacervate droplet).

Although the analytical treatment has implications for enzymes adsorbed or covalent bound to surfaces, or entrapped within gels, capsular or matrices, with the proviso that an unstirred layer prevails in the vicinity of the catalytic site, care is needed in micro-scopic systems such as with enzymes in biological membranes. However, the hetero-geneous distribution of substrate and product within the domain of enzyme particles is an important consideration in most circumstances.

Covalent binding of hexokinase and glucose-6-phosphate dehydrogenase to either sepharose or acrylic acid-acrylamide copolymers has confirmed the increased efficiency of such a spatially concentrated enzyme system over the equivalent homogeneous system [5]. Taking the rate of production of NADPH (Fig. 3) in the solution system as reference a 100–140% increase in the rate of NADPH production by the matrix bound enzymes was recorded. This could be qualitatively ascribed to the spatial proximity of the two enzymes on the matrix leading to higher local concentrations of the inter-mediate product, glucose-6-phosphate, than expected in solution. The two enzyme system was further extended to three enzymes by the same workers [6] by inclusion of β-galactosidase as a prior step in the sequence (Fig. 4). An immobilized system was pre-pared by the coupling of the three enzymes, via the CNBr method [12] to Sephadex G-50. The tightly crosslinked dextran gel was employed so that the complications arising

D-glucose + ATP $\xrightarrow{\text{hexokinase}}$ D-glucose-6-phosphate + ADP

D-glucose-6-phosphate + NADP$^+$ $\xrightarrow{\text{glucose-6-phosphate dehydrogenase}}$ D-gluconolactone-6-phosphate + NADPH

Fig. 3. Two step immobilized enzyme reactor of Mosbach and Mattiasson [5]

4-0-β-D-galactopyranosyl-D-glucose $\xrightarrow[\beta\text{-galactosidase}]{}$ D-galactose + D-glucose

D-glucose + ATP $\xrightarrow[\text{hexokinase}]{}$ D-glucose-6-phosphate + ADP

D-glucose-6-phosphate + NADP$^+$ $\xrightarrow[\substack{\text{glucose-6-phosphate}\\\text{dehydrogenase}}]{}$ D-gluconolactone-6-phosphate + NADPH

Fig. 4. Three step immoblized enzyme reactor described by Mattiasson and Mosbach [6]

from enzymes bound within the gel matrix would be avoided. The lag phase of the soluble
system was observed to be longer than that of the corresponding heterogeneous system
(i.e. the efficiency of the immobilized system was higher in the pre-steady state condition). The effect was more pronounced than in the two enzyme system. This was amply
illustrated by plots of $V_{\text{matrix}}/V_{\text{solution}}$ for the two and three enzyme systems. The
implication was, therefore, that on extending the number of enzymes in a consecutive
set of reactions and in a situation where the concentration of intermediates formed is
rate limiting, binding of the different enzymes to the same matrix had a accumulative
effect on the overall rate of reaction of the system in the initial phase. Such an effect
is not surprising, due to the increased local concentrations predicted in the microenvironment of the matrix bound enzymes. Whether the effect is caused by the proximity of
the bound enzymes to one another or the existence of a diffusion layer restricting
diffusion could not be elucidated but adsorbtion processes were ruled out as causing
increased local concentrations.

The theoretical treatment by Goldman and Katchalski [10] assumes an impermeable
membrane. The kinetics for a permeable membrane system, with the membrane permeable to both substrates and product, has been evaluated by Krishna and Rama-
chondran [13]. In addition to the above mentioned extension these authors examine
the case with relaxation of other criteria assumed by Goldman and Katchalski. Thus a
general analysis incorporating both inter- and intra-membrane (capsule) diffusional
resistances, spherical geometry in addition to the plate configuration, and without the
limit of constant bulk concentration, was devised whilst retaining a linear kinetic
scheme. The latter criteria is satisfied when enzyme concentrations E_1 and E_2 are constant, $K_{m1} \gg S$ and $K_{m2} \gg P_1$. With supposed reaction scheme

$$S \xrightarrow{k_1^*} P_1 \xrightarrow{k_2^*} P_2$$

it was shown that the effect of diffusion was not only to reduce the reaction rates but
also to change the observed reaction scheme. Diffusional resistances introduce an
apparent linkage between S and P_2, where none originally existed, with a disguised
rate constant k_3^*.

$$S \xrightarrow{k_3^*} P_2.$$

Physically this could be interpreted as indicating that measurement of changes in bulk
concentration and treating the system as homogeneous would give not only low kinetic

constants but also a mechanistic picture. Once the disguised rate constants are known the rates of reaction can be calculated from the derived equations. If the intrinsic first order rate constants k_1 and k_2 are known, and estimates of diffusion coefficient (D) and mass transfer coefficient (k_b) are available, the system performance may be predicted for a given geometry. Concentration profiles with time can be derived and plots for different conditions constructed. For short reaction times a large resistance by the membrane gives selectivity in favour of P_2. A small resistance by the membrane (low Thiele modulus $\emptyset_1$) favours P_1 production. Reduction of the k_2/k_1 ratio also favours P_1 production, with a lag period in P_2 production being observed if $k_1 > k_2$. In each case a maximum in P_1 production is observed, which was not revealed by the Goldman and Katchalski treatment due to the assumption that S remained constant. The treatment was extended to a packed bed two enzyme reactor by replacement of reaction time by a residence time in the packed bed giving analogous concentration profiles along the height of the bed.

First order reactions are characterized in this context by the independence of the "effectiveness factor" from substrate or product concentration in the bulk of solution. In cases where each step in the consecutive reaction follows Michaelis-Menten kinetics, with possible added complexities due to substrate or product inhibition, analytical expressions for the effectiveness factor cannot be obtained. This problem has been evaluated and the resultant non-linear differential equations solved by the orthogonal collocation method [4]. Calculation of the selectivity (net rate of formation of P_1 with respect to rate of formation of P_2) for various values of γ_1 (So/K_m for enzyme 1) showed that the selectivity is adversely affected by increase in γ_1. Conversely increasing γ_2 results in a considerable increase in selectivity due to suppression of reaction 2. The value of K_{m1} and K_{m2} (at constant $\gamma_1 = \gamma_2$) does not markedly alter selectivity although the effectiveness factor η (η is the ratio of the actual rate of reaction of S to the rate of reaction in the absence of diffusional, whether external or internal, effects) is increased. This approach was extended to cover the case of inhibition by products or substrates to give theoretical predictions of behaviour under various conditions. These results should prove useful in obtaining the optimum conditions for improving the rates and selectivity of consecutive reactions with respect to the design of enzyme reactors.

A detailed kinetic evaluation for a particular immobilized enzyme sequence has been performed using glucose oxidase and catalase [14]. A model system was defined as

$$A + G \rightarrow B + P$$

$$B \rightarrow nA$$

to parallel the proposed kinetic scheme for the double enzyme system. With the assumption that the first reaction was first order in A and zero order in G, and that the second was first order in B, justifiable by judicious choice of operating conditions, a mathematical representation of the steady state problem could be readily made. The results could be interpreted in terms of plots of catalytic activity versus the Thiele modulus for the first reaction for different reactivities and at differing surface concentrations. As the rate constant for the second reaction was increased the effect of internal diffusional restrictions were decreased. A maximum in catalytic activity for the first reaction was predicted with both reactants present.

This same system has been the subject of a more recent mathematical treatment with a parallel experimental study [15]. Glucose oxidase and catalase were immobilized in an inert carrier matrix of polyacrylamide gel. Concentration profiles with the particles were calculated under various limiting conditions.

In addition profiles of activity of glucose oxidase against time could be constructed against time and locality, taking into account the deactivation by the peroxide as shown by calculation. The observed experimental activity was in reasonable agreement with the theoretical prediction.

An important further consideration, in addition to the possible enhanced efficiency of the immobilized sequential enzyme system, is the improvement in rate of reaction where one step of the sequence normally has an unfavourable equilibrium. Since the distance between the sequential enzymes is decreased on matrix binding, and subsequently a higher diffusion rate caused by the steeper concentration gradient of the equilibrium, the overall reaction rate is improved. This has been demonstrated by Srere *et al.* [16] with the sequence malate dehydrogenase, citrate synthetase and lactate dehydrogenase.

The kinetic behaviour of an immobilized system subject to feed back inhibition has been studied and compared with a computer simulation [64]. Hydrolysis of salicin with β-glucosidase followed by oxidation with glucose oxidase, the enzymes being co-immobilized, results in the production of D-glucono-1,5-lactone which is an inhibitor of β-glucosidase. The agreement between experimental and theoretical data was good.

3 Specific Sequential Enzyme Systems

Glucose Oxidase and Catalase

The enzyme glucose oxidase (β-D-glucose: oxygen oxidoreductase: EC 1.1.3.4) specifically catalyses the oxidation of β-D-glucopyranose with the concomitant reduction of the FAD moiety of the enzyme to $FADH_2$. Subsequently oxygen is reduced to hydrogen peroxide. Hydrogen peroxide causes inactivation of glucose oxidase [17]. Catalase (H_2O_2 : H_2O_2 oxidoreductase EC 1.11.1.6) is responsible for the conversion of two molecules of hydrogen peroxide to one molecule of oxygen and water. The linked system glucose oxidase-catalase is therefore partially cyclic in character the product of reaction 2 being an effective cosubstrate for the first reaction. Since these two enzymes may be considered to be synergistic the simultaneous immobilization is particularly interesting.

An initial study of this sequential enzyme system was made by Messing [18] using the enzymes adsorbed into controlled pore size particles of alumina or titanium dioxide. Stable preparations of the two enzyme system were obtained provided that the pore size was greater than 35 nm (compared with a major axis of 18.3 for catalase and 8.4 for glucose oxidase, giving spinning diameters of ca. 36 and 17 nm respectively). Thus it appeared that only with preparations in which both enzymes were able to be retained within the support matrix pores was stability conferred. With pore sizes sufficient to exclude catalase but allow penetration by glucose oxidase, only low glucose oxidase apparent activites were observed, possibly due to the poor recycling of the peroxide by catalase resulting. Since peroxide causes inactivation of the glucose oxidase only suports with pores sufficiently large to allow catalase mediated breakdown of the peroxide produced gave stable preparations.

A systematic study of covalently bounded glucose oxidase and catalase on an inorganic support has been performed [19]. In this procedure nickelimpregnated silica alumina particles were coated with γ-aminopropyltriethoxy silane to produce a functionalised support. Coupling of enzymes was then effected by use of thiophosgene activation or glutaraldehyde cross linking. The controlling factor in enzyme immobilization was enzyme concentration, with competition between the enzymes for available binding sites being observed in the dual enzyme system.

This study was continued by the same workers to provide a comprehensive study of the relative efficiencies of soluble and immobilized glucose oxidase and catalase linked systems. Of particular interest was the additional direct comparison of particles to which both enzymes were immobilized with a mixed immobilized system where each enzyme was attached to different support surfaces. One advantage of studying a cyclic enzyme system is that either enzyme can be used as the first enzyme in the sequence depending on the substrate used. Calculation of lag times showed that, as expected, decreasing the rate constants for consecutive first order reactions gave increasing lag times. However, reversal of the rate constants between the two enzymes had no effect on lag times, thus the initiation with either enzyme should result in equivalent lag times for the overall sequential reaction. The measured efficiency of the dual immobilized system was superior to that of a soluble system with comparable activities as has been previously predicted. Generally the soluble system exhibited greater efficiencies than the mixed immobilized system, presumably due to necessity for P_1 to diffuse out of the particle containing enzyme 1 and then into the particle containing enzyme 2. The prediction by Goldman and Katchalski [10] that with a fixed ratio of enzyme activities the total efficiency of the system would increase with increasing absolute activities was confirmed by this work. As the rate constant for the second enzyme was decreased relative to that of the first enzyme the difference in efficiency and net product formation between the dual immobilized and homogeneous system diminished. When the second enzyme activity was low, a high proportion of intermediate could escape into the bulk phase and due to the concentration gradient, would be slow to diffuse back into the particle. Thus the main advantage of the dual immobilized enzyme system was negated and similar results to the soluble system are obtained. In general the dual immobilized enzyme system would be favoured by use of high activities and large particles with small pore sizes.

Glucose oxidase containing catalase has been immobilized on titanium dioxide particles (0.4–1.0 mm) impregnated with a copolymer of phenylenediamine and glutaraldehyde. The oxidation of glucose was then studied in a differential bed reactor [20]. Increase in the flow rate of glucose led to an increase in reaction rate for the total system. In the reactor the ratio of specific reaction rate in oxygen compared to air was dependent on flow rate and somewhat higher than in an ideal stirred system.

Deactivation occurred after between thirty and forty hours but addition of soluble
catalase restored the initial activity. Catalase is irreversibly denatured in the system.
A mathematical model in which the limitation by diffusion was neglected gave results
in reasonable agreement with those obtained experimentally.
The coupling reaction of glucose oxidase and peroxidase, for use in an immobilized
enzyme analytical system, to cyanogen bromide activated cellulose has been studied
in detail [21]. The coupling reaction was continuously monitored and the sequential
coupling of the two enzymes was shown to be complete within forty minutes.

β-Galactosidase and Glucose Oxidase
An interesting system of dual immobilized enzymes has been developed for the treatment of milk at its natural pH [22]. Milk contains a lacto peroxidase system for the
oxidation of thiocyanate by hydrogen peroxide, producing an intermediate capable
of killing many strains of gram-negative bacteria, but lacks the hydrogen peroxide or
a source of this compound on removal from the body. Since galactase oxidase
(EC 1.1.3.9) has a low activity towards lactose a two enzyme system using β-galactosidase (EC 3.2.1.23) and glucose oxidase (EC1.1.3.4) was used to generate hydrogen
peroxide from lactose via glucose. Use of *Aspergillus flavus* lactose was possible with
high activity in the coupled immobilized system at neutral pH in spite of the unfavourable pH optimum in solution for this enzyme. The combined system gave good bacteri-
cidal action in complex media when both enzymes were coupled simultaneously to
zirconium dioxide coated porous glass beads.

Starch Degrading Enzyme Systems
One of the major industrial applications of immobilized enzyme technology is in the
hydrolysis of starch. The use of multiple enzyme systems here is varied. For example
the conversion of starch to glucose using mixtures of amylolytic enzymes for efficient
reaction and the consequent conversion of glucose to fructose using glucose isomerase.

β-Amylase and Pullulanase
Maltose can be made from starch with β-amylase alone but is usually produced with
the aid of a mixture of β-amylase and α-amylase. With β-amylase alone a large portion
of β-limit dextrin from amylopectin is obtained whereas in the latter case a large amount
of glucose and low molecular weight branched dextrins are found. Use of an essentially
debranching enzyme such as pullulanase in conjunction with β-amylase has been reported
to give a good conversion of starch to a high maltose containing product [23]. Only low
amounts of glucose are formed arising solely from the chains with an odd number of
glucose units. Both pullulanase and β-amylase can be coupled separately to a crosslinked
copolymer of acrylamide and acrylic acid using a water soluble carboiimide [24–26].
This method of immobilization was used to attach both enzymes successively to the
above hydrophilic carrier to provide a two enzyme system for the production of high
maltose yields. This particular support offers the advantage of resistance to microbial
and enzyme degradation compared with polysaccharide based carriers.
The two enzyme system of β-amylase and pullulanase is of interest since it does not
carry out two consecutive reactions in the normal sense. Neither enzyme is capable of

converting starch completely without the aid of the other enzyme however. β-Amylase alone degrades the outer chains until the action is hindered by a branch point it cannot pass, whereas pullulanase in action alone is sterically hindered and consequently cannot split all the branching points. Hence simultaneous action is desirable, one enzyme opening the substrate structure for the other making complete hydrolytic degradation of the substrate possible.

The immobilized β-amylase/pullulanase preparation was considerably more stable than the corresponding soluble system. In a subsequent report the same workers extended the study of this system to include kinetic comparisons with the soluble enzyme system and to compare the practical and theoretical efficiencies of continuous stirred tank and plug-flow reactors. Although a full kinetic analysis is difficult due to the nature of the two reactions involved the system was simplified by choice of conditions where β-amylase activity could be regarded as the rate determining component. The apparent Michaelis constant, K_m', for the immobilized preparation was significantly higher than that for the corresponding free enzyme system presumably as a result of intra-particle diffusional resistance or steric hinderance of the substrate to enter the matrix pores. The longer denaturation half-life found at high substrate concentrations could be explained on the basis of the active site stabilising effect of the substrate.

Maltose is known to be a competitive inhibitor for β-amylase [27] and a measured apparent inhibitor constant, K_c', was obtained for the dual enzyme system. K_c' was lower than K_m' whereas in the free enzyme system K_i was greater than K_m. Theoretical expressions for the conversion of starch to maltose with time did not give a good correlation with experimental data probably due to the variation in β-amylase affinity for different molecular weight substrates.

As expected from the kinetic data obtained, when the effect of product inhibition and decreasing rate constant with hydrolysis extent is allowed for, a plug-flow reactor was theoretically and experimentally superior to a continuously stirred tank reactor. However, practical considerations of plugging and retrogradation may limit the use of a plug-flow reactor.

α-Amylase and β-Amylase

The study of the performance of a membrane restricted β-amylase in the hydrolysis of starch [28] was extended to the situation with a mixture of α- and β-amylases by Tachauer *et al.* [29]. In this work a mixture of barley

$$G(\alpha 1 \to 4)G_n(\alpha 1 \to 4)G \xrightarrow{\beta\text{-amylase}} G(\alpha 1 \to 4)G + G_{n-1}(\alpha 1 \to 4)G$$

$$G(\alpha 1 \to 4)G_n(\alpha 1 \to 4)G \xrightarrow{\alpha\text{-amylase}} G(\alpha 1 \to 4)G_x(\alpha 1 \to 4)G + G(\alpha 1 \to 4)G_y(\alpha 1 \to 4)G$$

malt α- and β-amylase (molecular weights of 60,000 and 197,000 respectively) were retained by a membrane of 18,000 molecular weight cut-off. This system was felt to be more relevent to the industrial hydrolysis of starch and leads to a mixture of short chain oligosaccharides. The membrane system in which the smaller molecular weight products of hydrolysis are continuously removed from the reaction solution gives a better performance than an equivalent solid wall reactor only at large reaction times. It was concluded that a sharper molecular weight cut-off was required to improve the performance of the membrane reactor at low reaction times. This experimental system has been the subject of a detailed theoretical study [30] highlighting the parameter effecting the reactor performance.

Glucoamylase and α-Amylase (and/or pullulanase)
Of considerable industrial significance is the use of amylolytic enzymes for the production of glucose from starch. For ultimate production of glucose

$$(G(\alpha 1 \rightarrow 4)G_n(\alpha 1 \rightarrow 4)G \xrightarrow{\text{glucoamylase}} G + G(\alpha 1 \rightarrow 4)G_{n-1}(\alpha 1 \rightarrow 4)G$$

the exoenzyme glucoamylase (γ-amylase or amyloglucosidase is required). In order to provide more chain ends for glucoamylase action α-amylase coaction is desirable. Pullulanase can also be beneficial in providing branchpoint cleavage at earlier stages of the reaction.
An immobilized system has been described with a combination of these enzymes [31]. Glucoamylase (ex *Aspergillus awamori*) was immobilized on DEAE-cellulose, and α-amylase (ex *Bacillus subtilis, Aspergillus oryzae*, or pancreatic) was bound to an activated aminoethylcellulose. With this system the major advantage is reflected in the increased yield of glucose resulting with the use of a combined enzyme system. Thus a partially hydrolysed starch (D. E. 16.9) under comparable conditions gave 71.5% conversion to glucose with glucoamylase and between 89.0 and 93.1% with a mixed enzyme system. A more rapid production rate of glucose would be anticipated with a combined enzyme system since a higher local concentration of potential substrate for glucoamylase action would ensue.

Glucoamylase and Glucose Isomerase
In view of the recent interest in the production of high fructose syrups the combination of a glucose isomerase (E.C. 5.3.1.5) with a starch hydrolysis enzyme system is of particular interest.

$$G(\alpha 1 \rightarrow 4)G_n(\alpha 1 \rightarrow 4)G \xrightarrow{\text{glucoamylase}} G_{n+2}$$

$$\text{D-Glucose} \xrightleftharpoons{\text{glucose isomerase}} \text{D-fructose}$$

Although the normal equilibrium position for glucose isomerase gives an equimolar proportion of D-glucose and D-fructose use of oxyanion to displace the equilibrium in favour of D-fructose can be used with advantage. Both borate [32] and germanate [33] may be used in this connection.
Glucoamylase and glucose isomerase have been co-immobilized on porous glass particles [34]. Arrhenius plots and pH profiles have been determined for both enzymes. The stability of glucoamylase was linearly dependent on starch concentration, as a result of substrate protection, whereas the glucose isomerase stability showed a sigmoidal dependence on fructose concentration. The combined enzyme bed was operable for at least one month at pH 6.0–6.5 and 55°.

Since many glucose isomerase preparations consist of cell aggregates it would be of considerable interest to examine the characteristics of such cell aggregates with a starch hydrolysing enzyme such as glucoamylase attached to the outer surface of the aggregate.

Glucoamylase and Glucose Oxidase
Maltose has been used as a model substrate for the production of gluconic acid from starch using a combined enzyme system with glucoamylase and

$$\text{Maltose} \xrightarrow{\text{glucoamylase}} 2 \text{ D-glucose}$$

$$\beta\text{-D-glucopyranose} + O_2 \xrightarrow{\text{glucose oxidase}} \text{D-glucono-1,5-lactone} + H_2O_2$$

glucose oxidase. This pair of enzymes illustrates the advantages gained from immobilization when two enzymes have incompatible pH optima. Glucoamylase (pH optimum 4.8) and glucose oxidase (pH optimum 6.4) coimmobilized on sepharose, via cyanogenbromide activation, show a difference between the pH optima of the coupled reaction and the soluble system of 0.3 pH units [35]. The relative position of the pH optimum of such a system is dependent upon the activity quotient of the two enzymes.
Thus a difference of up to 0.75 pH units could be obtained by variation in the activity quotient. With a large excess of glucose oxidase, the first reaction catalysed by glucoamylase becomes rate limiting and the pH dependence of the combined reaction follows that of the first reaction in the sequence. A decrease in the activity of glucose oxidase leads to an increase in the pH optimum of both the soluble and immobilized system. However, due to diffusional limitations giving rise to higher local concentrations of glucose in the matrix, an induced higher activity of glucose oxidase would be apparent in the immobilized system and results in a larger shift in the overall pH optimum towards that of the glucose oxidase.
A further complication in this system may arise from hydrolysis of the end product, D-glucono-1,5-lactone, to D-gluconis acid with a resultant pH change. Thus the local pH microenvironment may be affected by the overall reaction rate and hence gives rise to a modified pH profile at different reaction rates. This combination of enzymes, immobilized on glass fibres, has been used in conjunction with an oxygen electrode to provide an electrode responding to starch [36].

β-Glucosidase and Glucose Oxidase
Although not strictly an immobilized multienzyme system the report [37] of a soluble aggregate of two enzymes is of interest. Coupling of almond

$$\text{p-Nitrophenyl-}\beta\text{-D-glucopyranoside} \xrightarrow{\beta\text{-glucosidase}} \text{p-nitrophenol} + \text{D-glucose}$$

$$\text{D-Glucose} + O_2 \xrightarrow{\text{glucose oxidase}} \text{D-glucono-1,5-lactone} + H_2O_2$$

β-glucosidase and glucose oxidase (ex *Aspergillus niger*) was effected by crosslinking with glutaraldehyde to give a bifunctional, soluble, aggregate with molecular weights in the range $2 \times 10^5 - 3 \times 10^5$. This system provides a useful model for kinetic data with linked enzyme systems due to the minimization of effects such as unstirred layers. Thus a much closer examination of proximity effects may be feasible. In contrast to the matrix-bound systems no pronounced lag phases between aggregates and free enzymes were observed. It is not clear however whether any particular advantages accrue from the aggregation of the two enzymes with respect to the individual enzymes.

Invertase and Glucose Oxidase
These two enzymes have been coimmobilized to the inner surface of nylon tubes for the analysis of sucrose [43]. No advantage accrued from

$$\text{Sucrose} \xrightarrow{\text{invertase}} \text{D-glucose} + \text{D-fructose}$$

$$\text{D-glucose} + O_2 \xrightarrow{\text{glucose oxidase}} \text{D-glucono-1,5-lactone} + H_2O_2$$

coimmobilization, compared with separate immobilized enzymes. This observation has been explained in terms of the slow mutarotation of α-D-glucopyranose (the product of invertase action on sucrose) into β-D-glucopyranose, the substrate for glucose oxidase, being rate limiting.
A similar system employing invertase and glucose oxidase bound to glass fibres has been used in conjunction with an oxygen electrode to provide an electrode responsive to sucrose [36].

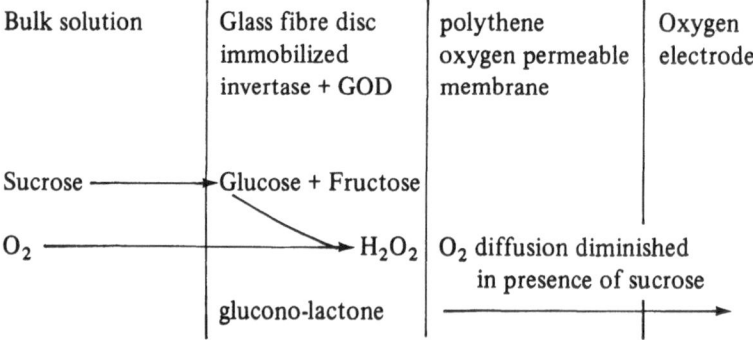

Phosphorylase a, Phosphoglucomutase, Phosphoglucose isomerase, and Phosphatase
This enzyme sequence has been studied using enzymes covalently bound to cellulose, in conjunction with anionic and neutral membranes, for the conversion of starch or glycogen into fructose [39]. Anionic membranes were employed to prevent migration of phosphated intermediate species from the reaction system. The enzymes in this example were always separately immobilized to cellulose, the phosphorylase a being coupled via diazotised 3-(-p-amino-phenoxy)-2-hydroxypropyl cellulose whilst phosphoglucomutase, phosphoglucose isomerase and phosphatase were coupled to cyanogen bromide activated cellulose. The individual enzyme reactions were initially optimised, with due reference to the compatibility of the different enzyme requirements, before examination of sequential reactions. Ultimately the four enzyme sequence was

	A	B	C	D	E	F	buffer
Glycogen G-1,6-DP							

A : immobilized phosphorylase a; B immobilized phosphoglucomutase; C neutral membrane; D phosphoglucose isomerase; E phosphatase; F anionic membrane.

Glycogen – A → G-1-P – B → G-6-P – D → F-6-P – E → Fructose

Diagram 1. Schematic diffusion cell for the sequence glycogen → fructose

examined using a diffusion cell constructed as shown in Diagram 1. Analysis of the cell compartments with time indicated that no significant loss of phosphate esters had occured from the substrate chamber even after 60 hours operation. In spite of the use of borate to influence the phosphoglucose isomerase reaction in favour of fructose, the product stream contained significant quantities of glucose although less than would be expected from the normal equilibrium position of the enzyme. In view of the non specificity of the phosphatase, selective production of fructose could only be achieved if the glucose-6-phosphate concentration was low. Considerable problems were experienced with the use of borate since glucose-1-phosphate-borate complex was an effective inhibitor of phosphoglucomutase action.

Hexokinase, Glucose-6-Phosphate Dehydrogenase, β-Galactosidase, Hexokinase and Glucose-6-Phosphate Dehydrogenase
This two enzyme [5] and three enzyme [6] sequence has been mentioned in an earlier section.

Hexokinase and Phosphatase

$$\text{D-Glucose} + \text{ATP} \xrightarrow{\text{hexokinase}} \text{D-glucose-6-phosphate} + \text{ADP}$$

$$\text{D-glucose-6-phosphate} \xrightarrow{\text{phosphatase}} \text{D-glucose} + \text{inorganic phosphate.}$$

These two enzymes have been immobilized in a structured multilayer bienzyme membrane to demonstrate active transport of glucose against a concentration gradient [40, 41]. The membrane used comprised two protein layers, hexokinase and phosphatase covered on their outer surfaces with selective membranes permeable to glucose rather than glucose-6-phosphate. Permeation of glucose into the sandwich results initially in phosphorylation by hexokinase and ATP to glucose-6-phosphate. The glucose-6-phosphate permeates into the phosphatase layer and is consequently hydrolysed to glucose before diffusion out of the sandwich can occur. Since glucose is a substrate of the first enzyme layer but a product of the second enzyme layer a sinusoidal glucose concentration profile exists

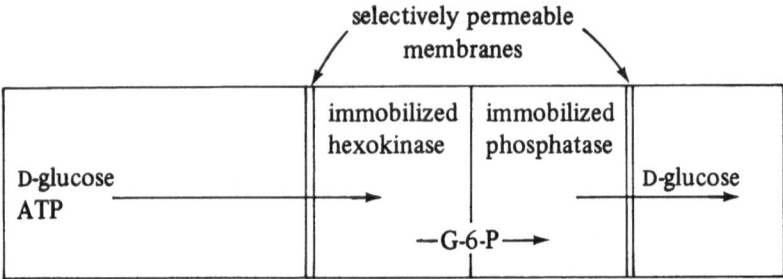

across the membrane. Hence a differential glucose concentration exists at the two
surfaces of the membrane sandwich causing glucose to enter at the hexokinase side
and leave at the phosphotase side of the sandwich. The net effect of the system is thus
to achieve transport across a spatial and enzyme sequential array driven by ATP supplied
to the system.

Hexokinase, Phosphoglucose Isomerase, Phosphofructokinase, and Aldolase

This sequence of glycolytic enzymes has been studied [7] using enzymes immobilized
on separate particles.

Glucose → G-6-P ⇌ F-6-P → F-1,6-DP → glyceraldehyde-3-P +
dihydroxyacetone phosphate

Malate Dehydrogenase, Citrate Synthase

These enzymes are interesting since the first reaction, that of malate to oxalacetate is
reversible, and, moreover thermodynamically unfavourable for oxalacetate formation.
Coimmobilization of malate dehydrogenase and citrate synthase resulted in a displace-
ment of this equilibrium with a two-fold increase in the steady state rate of citrate formation compared with the solution reaction of the two enzymes [16].
These enzymes and lactate dehydrogenase were coupled to cross-linked dextran and
sepharose and used as a model for microenvironment compartmentation in mitochondria.
Lactate dehydrogenase was used to mimic the

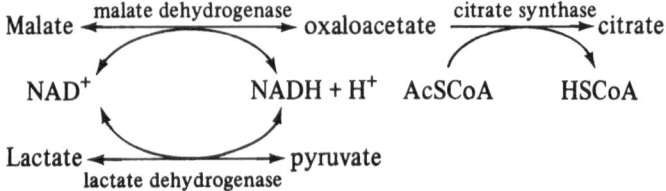

DPNH utilising system of mitochondria, and the rate of citrate production from malate,
DPN and acetyl coenzyme A was determined continuously in a flow system. Up to 100%
enhancements in rate were observed with the immobilized system compared with the
soluble system, and 400% increases in reaction rate were observed on addition of pyruvate to reoxidise the DPNH formed.

An aggregate of malate dehydrogenase and citrate synthase has also been described [37].

Pyruvate Kinase and Lactate Dehydrogenase
These enzymes have been immobilized separately on cellulose (filter paper) and used in a two enzyme continuous feed reactor [42]. These two enzymes have also been immobilized by binding to diazotised derivatives of glass beads and used consecutively to determine pyruvate and phosphoenol pyruvate [38].

Tryptophanase and Lactate Dehydrogenase
Coimmobilization of these enzymes to sepharose, using cyanogen bromide activation, has been shown to be advantageous in the analysis of L-tryptophan [44].

$$\text{L-Tryptophan} \rightleftharpoons \text{Indole} + NH_4^+ + \text{pyruvate}$$

$$\text{pyruvate} + NADH + H^+ \rightarrow \text{lactate} + NAD^+.$$

The proximity between the reacting enzymes results in a more rapid response initially and importantly a higher sensitivity towards low concentrations of L-tryptophan.

Aspartate Amino Transferase and Malate Dehydrogenase
Similarly proximity effects have been suggested as responsible for the 10 fold increase in sensitivity of the coimmobilized system compared with a reaction scheme wherein aspartate amino transferase was immobilized and malate dehydrogenase was used in solution [45].

$$\text{L-Aspartate} \rightarrow \text{oxaloacetate} + NH_4^+$$

$$\text{oxaloacetate} + NADH + H^+ \rightarrow \text{malate} + NAD^+.$$

Trypsin, Glucose Oxidase and Hexokinase
Entrapment of these three non consecutive enzymes was effected in polyacrylamide gel to enable the study of a model for substrate flow through different metabolic pathways [46]. Thus hexokinase and glucose oxidase can be in competition for the same substrate. At pH 8.5 the proportion of glucose phosphorylated by hexokinase (pH 8.5 optimum) decreased when trypsin was acting on benzoylarginine ethyl ester. This was compensated by an increase in the oxidation by glucose oxidase (pH 6.6 optimum) during the period of trypsin action. The initial state was restored on cessation of trypsin activity. The generation of protons on trypsin action was sufficient in the gel matrix to alter the balance between the two glucose consuming enzymes with their different pH optima.

Regeneration of Coenzymes
An intriguing use of successive immobilized enzymes has been the development of systems for the regeneration *in situ* of coenzymes, a necessary prerequiste for the use of many enzyme systems in an economic manner.

Use of adenylate kinase (AMP:ATP phosphotransferase, E. C. 2.7.4.3) and acetate kinase (ATP:acetate phosphotransferase, E. C. 2.7.2.1) have been utilised to produce an immobilized enzyme system for the continued regeneration of ATP from AMP or ADP [47]. The ultimate phosphorylating agent is acetylphosphate, readily obtainable from acetone. The two enzymes were immobilized by entrapment in a polyacrylamide gel, the process requiring careful control to avoid inactivation of the enzymes, both of which contain structurally

$$ AMP + ATP \xrightarrow[\text{kinase}]{\text{adenylate}} 2\ ADP;\ 2\ ADP + 2\ AcP \xrightarrow[\text{kinase}]{\text{acetate}} 2\ ATP + 2\ Ac $$

important and reactive cysteine residues near their active sites, by the acrylic monomers. The result gel system provides a very efficient process of ATP regeneration. Theoretically a reactor containing acetate kinase (30 mg, 300 IV specific activity) and adenylate kinase (2 mg, 2000 IV specific activity) is capable of regenerating a kilogram of ATP per day. A more studied system is the recycling of NAD^+ to NADH or *vice versa*. Thus a triple enzyme system has been shown to function when the enzymes

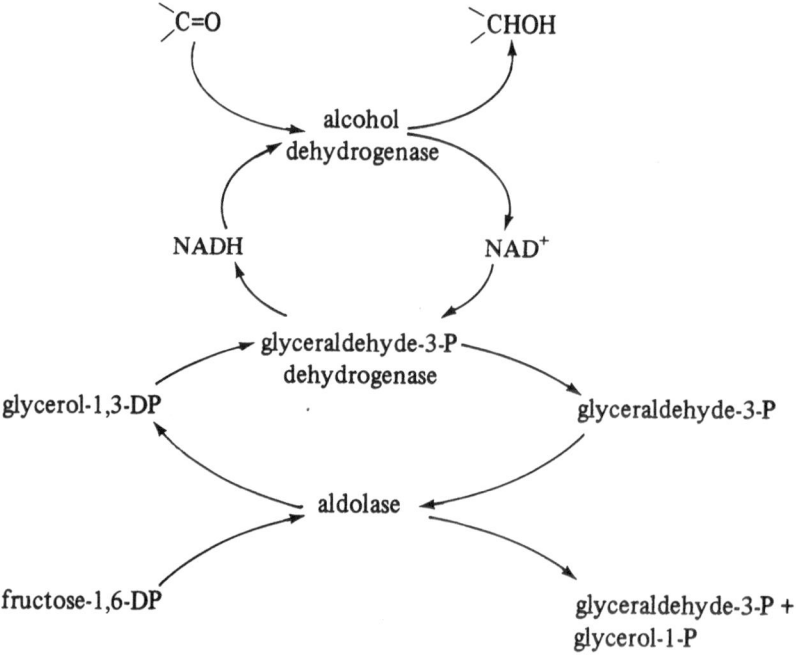

are immobilized [48] but as yet the NADH generating capacity of this system has not been fully utilised for carbonyl reduction on a large scale.

A further viable system for NAD^+ regeneration has been demonstrated in which the enzymes are pumped through hollow fibres surrounded by the reaction mixture [49].

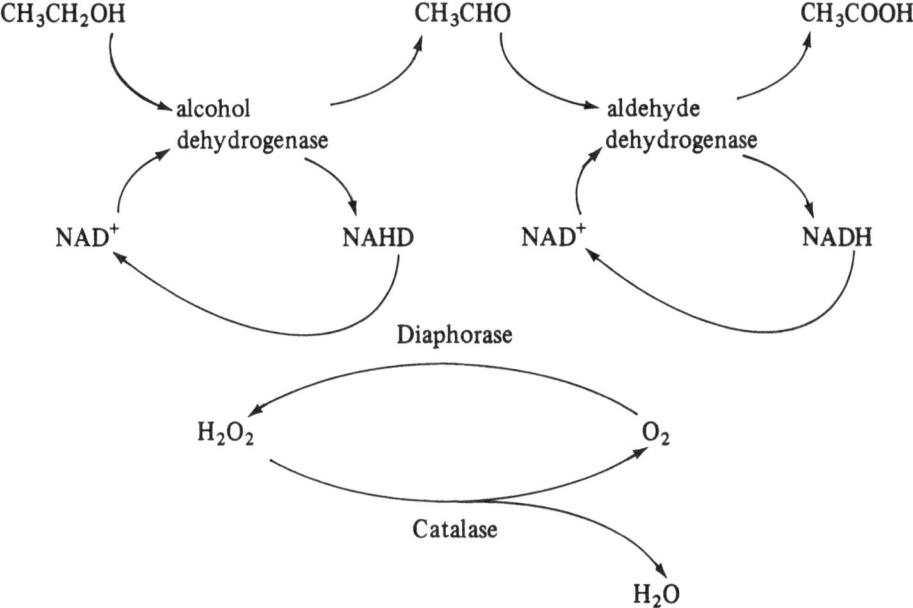

The semipermeable walls of the fibres allow passage of substrates, products and co-enzymes but not the enzymes. The enzymes are thus available for re-use on removal of the fibres from the reaction medium. Although specifically used for a study of oxidation of ethanol to acetic acid, the principle involved is of more general application.

4 Immobilized Whole Cells

Considerable interest has arisen in recent years in the use of whole cells, either entrapped in matrices, bound to particles, or as aggregates, for the production of various materials. The use of whole cells often obviates the difficulty of enzyme isolation and sometimes overcomes the problems of enzyme instability. There can often be more specific advantages when a given process requires a number of enzyme reactions, as the cellular enzyme systems will perhaps contain the enzymes organised in the requisite pathways. Where coenzymes are required, the immobilized cell will normally be able to recycle these as a matter of course thus greatly simplifying the operation of the system and giving an added cost benefit. Such systems are obviously not without their own disadvantages particularly if the cells are to remain viable.

Gel entrapment has been considerably exploited for the immobilization of whole cells and has been exploited for the production of organic acids including amino acids [50–53], the interconversion of carbohydrates [54–56] and steroid transformation [51]. In one of the first examples of this technique, lichen cells were immobilized within a crosslinked polyacrylamide gel and used to perform ester hydrolysis and acid decarboxylations in a column reactor. This system retained a part of the esterase activity and decarboxylase activity after three months at $20°$. Chibata [50, 51] and his coworkers have used several polyacrylamide immobilized aspartase containing microorganisms for the production of L-aspartic acid from ammonium fumarate.

$$HOOC - CH = CH - COOH + NH_3 \rightarrow HOOC - CH_2 - \underset{\underset{NH_2}{|}}{CH} - COOH$$

In addition these workers have used immobilized cells for the production of L-citrulline [52], urocanic acid [53] and 6-aminopenillanic acid [58]. Thus L-citrulline was produced from L-arginine by the L-arginine deaminase from *Pseudomonas putida,* the system having a half-life of 140 days at $37°$. Similarly *Achromobacter liquidum,* immobilized in polyacrylamide gel, gave high yields of urocanic acid form L-histidine with a half-life of 180 days. The volume of medium required for the production of a given quantity of product is smaller with a continuous method with immobilized cells than with a conventional submerse culture.

A two stage process for the conversion of Reichsteins compound S, (pregna-4-ene-17α, 21-diol-3,20 dione) via cortisol, into prednisolone has been described [57]. The initial reactions involving 11-β-hydroxylation was carried out by gel entrapped *Curvularia lunata* mycelia, and the subsequent 3-keto steroid-Δ^1-Dehydrogenase catalysed reaction by gel entrapped *Arthobacter simplex.* An important feature of this system was that complete transformation was possible, thus avoiding the complication of product purification. Further the system was interesting in that the system after exhaustion was capable of reactivation by treatment with nutrients and a steroid inducer *in situ.* The 3-keto-steroid-Δ^1-dehydrogenase activity was approximately ten fold higher with immobilized cells compared with enzyme isolated from the same weight of cells. Thus the extraction process not only would be time consuming but also less efficient.

Collagen has been used as a matrix for immobilization of microbial cells by Vieth and his coworkers [54]. For example, heat treated cells of *Streptomyces phaeochromogenes,* containing glucose isomerase, have been immobilized on hide collagen in the form of a membrane and demonstrated a high retention of activity at 70 °C. The increased thermal stability of this system over glucose isomerase itself is vitally important for the commercial application of glucose isomerase for the production of high fructose syrups. Glucose isomerase has also been conveniently immobilized by flocculation using a poly vinylimidazoline to give a stable preparation [59]. Cell agglomerates of this type have also been used for the inversion of sucrose and the specific deacylation on an acylated amino acid racemate [60].

It has been suggested that the use of whole cells in columns is not ideal since they will tend to grow and reproduce, and therefore those spores which are in the resting stage are more suitable for use. These are not physiologically inert and are capable of selective enzymic activity. Thus spores of *Aspergillus* and *Penicillium* species have been immobilized in columns of ECTEOLA-cellulose and shown to exhibit invertase activity [61].

It is perhaps surprising that there are relatively few publications in this field in view of the benefits that can accrue. An impressive example is the system involving *Brevibacterium ammoiagenes*, immobilized on a polyacrylamide gel, for the production of the pharmaceutically important L-malic acid from sodium fumarate. It has been calculated that a 1000 litre column of these immobilized cells, with a feed of $200\ \mathrm{l/h^{-1}}$ could satisfy in one month the annual demand for this compound [62].

5 Conclusion

It is a general feature of metabolic processes that the product of one enzyme reaction is the substrate for a subsequent reaction step. Furthermore there is now a wealth of evidence to suggest that the various cellular enzymes are not isolated entities but exist as aggregates, or are membrane bound. For example it has been elegantly demonstrated that in extracts of *Euglena* cells, practically all the enzymes were in various strata of the cell, leading to the conclusion that the overwhelming majority of enzymes in the cell were bound in some manner to sub-cellular matrices [63]. The continued study of immobilized multiple enzymes systems is thus important for a knowledge of the mode of enzymes in biochemical situations. It is probable however that from an industrial standpoint the major area of interest will be the exploitation of immobilized whole cells.

6 References

1. Bar-Eli, A., Katchalski, E.: J. Biol. Chem. **238**, 1690 (1963)
2. Lilly, M. D., Hornby, W. E., Crook, E. M.: Biochemical J. **100**, 718 (1966)
3. Weetall, H. H., Havewala, N. B.: In Enzyme Engineering (Biotechnol. Bioeng. Symp. No. 3), Wingard, L. B. (ed.). New York: Wiley-Interscience, 1972, p. 241
4. Ramachandran, P. A., Krishna, R., Panchal, C. B.: J. Appl. Chem. Biotechnol. **26**, 214 (1976)
5. Mosbach, K., Mattiasson, B.: Acta Chem. Scand. **24**, 2093 (1970)
6. Mattiasson, B., Mosbach, K.: Biochim. Biophys. Acta **235**, 253 (1971)
7. Brown, H. D., Patel, A. B., Chattopadhyay, S. K.: J. Chromatog. **35**, 103 (1968)

8. Kuchel, P. W., Nichol, L. W., Jeffrey, P. D.: J. Theor. Biol. **48**, 39 (1974)
9. Easterby, J. S.: Biochim. Biophys. Acta **293**, 552 (1973)
10. Goldman, R., Katchalski, E.: J. Theor. Biol. **32**, 243 (1971)
11. Vasitera, N. V., Balaevskaya, T. O., Gogilashvili, L. Z., Serebrovskaya, K. B.: Biochemistry **34**, 641 (1969)
12. Axen, R., Porath, J., Ernback, S.: Nature **241**, 1302 (1967)
13. Krishna, R., Ramachandran, P. A.: J. Appl. Chem. Biotechnol. **25**, 623 (1975)
14. Lawrence, R. L., Okay, V.: Biotechnol. Bioeng. **15**, 217 (1973)
15. Buchholz, K., Reuss, M.: Chimica **31**, 27 (1977)
16. Srere, P. A., Mattiasson, B., Mosbach, K.: Proc. Natl. Acad. Sci **70**, 2534 (1973)
17. Kleppe, K.: Biochemistry **5**, 139 (1966)
18. Messing, R. A.: Biotechnol. Bioeng. **16**, 897 (1974)
19. Bouin, J. C., Atallah, M. T., Hulkin, H. O.: Biochim. Biophys. Acta **438**, 23 (1976)
20. Prenosil, J. E., Carter, R., Dunn, I. J.: Chimica **31**, 109 (1977)
21. Cremonesi, P., Mazzola, G.: Cell. Chem. Technol. **10**, 567 (1976)
22. Björck, L., Rosen, C. G.: Biotechnol. Bioeng. **18**, 1463 (1976)
23. Ger. Offen. 1916741 (1970); 1958014 (1970)
24. Martensson, K.: Biotechnol. Bioeng. **16**, 567 (1974)
25. Martensson, K.: Biotechnol. Bioeng. **16**, 579 (1974)
26. Martensson, K.: Biotechnol. Bioeng. **16**, 1567 (1974)
27. Misra, V. K., French, D.: Biochem. J. **77**, IP (1960)
28. Closset, G. P., Cobb, J. T., Shah, Y. T.: Biotechnol. Bioeng. **16**, 345 (1974)
29. Tachauer, E., Cobb, J. T., Shah, Y. T.: Biotechnol. Bioeng. **16**, 545 (1974)
30. Subramanian, T. V.: Biotechnol. Bioeng. **18**, 1473 (1976)
31. Ger. Offen. 2538322 (1976)
32. Takasaki, Y.: Agr. Biol. Chem. **35**, 1371 (1971)
33. Barker, S. A., Somers, P. J., Woodbury, R., Stafford, G. H.: Brit. Pat. 1497888 (1978)
34. Lee, G. K.: Rep.-Kans. State Univ., Inst. Syst. Des. Optim. **66**, 32 (1975)
35. Gestrelius, S., Mattiasson, B., Mosbach, K.: Biochim. Biophys. Acta **276**, 339 (1972)
36. Barker, S. A., Somers, P. J., Eagling, P. R. E.: British Patent applied for.
37. Mattiasson, B., Johansson, A. C., Mosbach, K.: Eur. J. Biochem. **46**, 431 (1974)
38. Newrith, T. L., Diegelman, M. A., Pye, E. K., Kallen, R. G.: Biotechnol. Bioeng. **15**, 1089 (1973)
39. Rattle, S. J.: Ph. D. Thesis, University of Birmingham (1975)
40. Thomas, D., Bourdillon, D., Broun, G., Kernevez, J. P.: Biochemistry **13**, 2995 (1974)
41. Broun, G., Thomas, D., Selagny, E.: J. Membr. Biol. **8**, 313 (1972)
42. Wilson, R. J. H., Kay, G., Lilly, M. D.: Biochem. J. **109**, 137 (1968)
43. Inman, J. K., Hornby, W. E.: Biochem. J. **137**, 25 (1974)
44. Ikeda, S., Fukui, S.: FEBS Lett. **41**, 216 (1974)
45. Ikeda, S., Sumi, Y., Fukui, S.: FEBS Lett. **47**, 295 (1974)
46. Gestrelius, S., Mattiasson, B., Mosbach, K.: Eur. J. Biochem. **36**, 89 (1973)
47. Whitesides, G. M., Lamotte, A., Adalsteinsson, D., Colton, G. K.: Methods in Enzymol. **44**, 887 (1976)
48. Falb, R. D., Lynn, L., Shapira, J.: Experientia **29**, 958 (1973)
49. Chambers, R. P., Ford, J. R., Allender, J. H., Baricos, W. H., Cohen, W.: In Enzyme Engineering, 2, Pye, E. K., Wingard, L. B., Eds. New York: Plenum Press, 1973, pp. 195–202
50. Chibata, I., Tosa, T., Sato, T.: Appl. Microbiol. **23**, 878 (1974)
51. Tosa, T., Sato, T., Mori, T., Chibata, I.: Appl. Microbiol. **27**, 886 (1974)
52. Yamamoto, K., Sato, T., Tosa, T., Chibata, I.: Biotechnol. Bioeng. **16**, 1589 (1974)
53. Yamamoto, K., Sato, T., Tosa, T., Chibata, I.: Biotechnol. Bioeng. **16**, 1601 (1974)
54. Vieth, W. R., Wang, S. S., Saini, R.: Biotechnol. Bioeng. **15**, 565 (1973)
55. Saini, R., Vieth, W. R.: J. Appl. Chem. Biotechnol. **25**, 115 (1975)
56. Toda, K., Shoda, M.: Biotechnol. Bioeng. **17**, 481 (1975)
57. Mosbach, K., Lersson, P. O.: Biotechnol. Bioeng. **12**, 19 (1970)

58. Sato, T., Tosa, T., Chibata, I.: Eur. J. Appl. Microbiol. **2**, 153 (1976)
59. U. S. Patent 161337 (1971)
60. Lee, C. K., Long, M. E.: U. S. Patent 3821086
61. Johnson, D. E., Ciegler, A.: Arch. Biochem. Biophys. **130**, 384 (1969)
62. McAbee, M. K.: Chem. Eng. News **53**, 32 (1975)
63. Kempner, E. S., Miller, J. H.: Exptl. Cell Res. **51**, 150 (1968)
64. Lecoq, D., Hervagault, J. F., Broun, G., Joly, G., Kernevez, J. P., Thomas, D.: J. Biol. Chem. **250**, 5496 (1975)

Carriers for Immobilized Biologically Active Systems

R. A. Messing
Corning Glass Works
Fundamental Life Sciences, Biomedical and Chemical Technology
Corning, NY 14830, USA

Contents

1 Introduction 52
2 Surface Contributions 53
   2.1 Microenvironmental Effects 53
       2.1.1 pH, Surface pK, and Buffering Effects 53
       2.1.2 Hydrophobic and Hydrophilic Carrier Surfaces 56
       2.1.3 Redox Surfaces 56
       2.1.4 Metal Ions 57
   2.2 Side Reactions 57
       2.2.1 Non-Specific Adsorption or Accumulations 57
       2.2.2 Blood Clotting Reaction and Immune Response 58
   2.3 Surface Stability or Durability 58
3 Carrier Properties, Morphology, and Configuration 59
   3.1 Non-Porous Carrier 60
   3.2 Porous Carriers 60
   3.3 Gel Structure Carriers 61
   3.4 Rigid Versus Elastic Carriers 61
   3.5 Microbial Considerations 62
   3.6 Thermal Stability of Carrier 63
   3.7 Carrier Shape and Practicle Size 63
   3.8 Preliminary Carrier Selection 63
   3.9 Optimizing a Dimensionally Stable Controlled-Pore Carrier 63
4 Modification and Preconditioning of Carriers Prior to Immobilization 67
5 Regeneration of Systems 68
   5.1 Reactivation of the Immobilized Enzyme 68
   5.2 Regeneration of Immobilized Enzyme 68
   5.3 Regeneration of the Carrier 68
6 Methods for Immobilizing by Adsorption 69
   6.1 The Static Procedure 69
   6.2 Electrodeposition 69
   6.3 Reactor Loading Process 69
   6.4 Mixing or Shaking Bath Loading 70
7 Approaches to Compensation for Surface Contributions 70
   7.1 Zirconia Surface Treatment of Glass for Durability 70
   7.2 Stannous Treatment of Titania Surfaces 71
   7.3 Aromatic Surfaces to Increase Durability 71
8 Concluding Remarks and Additional Precautions that Concern the Carrier and the
   Immobilized Systems 71
9 References 72

1 Introduction

An immobilized enzyme should be viewed as a system whose individual entities operate
in harmony such that no individual component counters the contribution of any of the
other components. In other words, the immobilized enzyme system, ideally speaking,
is a composite formed for the purpose of inducing the individual parts to perform in a
harmonized manner so that the composite output is either equal to, or greater, than the
sum of the output of the individual components. It can be claimed that the latter statement is qualitative in nature; however, that statement can be quantified by simply stating
that an immobilized enzyme system should be evaluated by the cost of the contribution
of the individual components versus the value of the performance of the system. With
this in mind, the prime concern is to identify the components of the immobilized enzyme
system.
Traditionally, the major components cited for an immobilized enzyme system have been
the enzyme, the carrier, and the mode of attaching the enzyme to the carrier. If these
were the only three entities that were considered in an evaluation program for an immobilized enzyme system, the technology would have been grossly over-simplified. Additional components which contribute to the environment and thus to the performance
of the enzyme must be considered. Included in these factors (components) are pH,
temperature, ionic strength, pressure, agitation, cofactor delivery, and substrate delivery
with product removal. All of these factors have a bearing upon the performance of the
carrier, and thus, inevitably affects the performance of the enzyme.
With the sole exception of the enzyme, the most important contributing component
to the performance of an immobilized enzyme system is the carrier. A carrier, judici-
ously chosen, can enhance the half-life of the immobilized system. A carrier impru-
dently chosen will adversely affect not only the half-life but also the performance of
the immobilized enzyme. The first point to recognize in the selection of a carrier is
that *there is no universal carrier*. A carrier which is a poor choice for one application
may be the best choice for another. The environment, the economics, and the specific
conditions of the application should dictate the choice of the carrier rather than the
prejudices of advocates or the apparent low cost of the commodity carrier.
A variety of carrier properties (i.e. surface contributions, morphology, configuration,
composition, preconditioning) should be assessed with respect to the specific application prior to the selection for evaluation. The properties of the carrier will inevitably
affect considerations with respect to regeneration. Thus, regeneration becomes an important factor which must be considered in the total economics of the immobilized
enzyme system. The purpose of this chapter is to indicate a rational approach for se-
lecting carriers based on their properties so that they may contribute synergistically
to the enzyme activity in a chosen environment in order to obtain an optimal economic performance.
Although enzymes employed in hollow fibers or contained within dialysis membranes
have in some instances been classified as immobilized, these will not be considered in
the discussions that will follow.

2 Surface Contributions

An enzyme is placed in intimate contact with the surface of a carrier by the immobilization process. The enzyme may be either attached directly to the surface by adsorptive forces or may be fixed more remote from the surface with the application of a spacer arm which is attached to the surface. The surface contributions are much greater when the enzyme is attached directly to the surface than when it is in a more remote situation. In the case of a spacer arm, the surface effects of the spacer may be as dramatic, or even more so, as that of the surface of the carrier.

Gross observations made in the very early studies of immobilized enzymes [1–3] clearly demonstrated that the surface of the carrier whether it be of inorganic or organic composition affected the activity of the enzyme. The environment of the enzyme attached to the surface of the carrier was found to be quite different than that of the bulk solution. The term "microenvironmental effects" was applied to that local environment in which the enzyme operated and which was dramatically affected by the carrier or coupling agent. This microenvironment may have a dramatic effect upon the apparent kinetics of the enzyme reaction; therefore, those factors that contribute to the microenvironment should be carefully analyzed with respect to the application. A broad spectrum of carriers that have become available allow an almost infinite choice of surface contributions.

Other than microenvironmental effects, two additional surface contributions play an important role in the effectiveness of a carrier for an enzyme application. The first and most important surface effect is that of stability or durability under the application conditions while the second effect may be termed side reactions. From these discussions, it becomes clearly apparent that the term "inert" should not be applied to a carrier. In the author's experience, all carriers appear to contribute some surface effect and since the term "inert" implies that no effect is contributed by the carrier, this terminology is not appropriate.

2.1 Microenvironmental Effects

Microenvironmental effects have two different dimensions. One of these dimensions includes the modification of the net charge on the surface of the enzyme protein. This is clearly apparent from a review [4] of the work of Abramson, Moyer, Gorin, and Tiselius [5–8]. These researchers demonstrated that the net charge on a protein surface as indicated by electrophoretic techniques was dramatically changed by the surface charge of an insoluble material. In other words, there was a charge-charge interaction, at times unpredictable, between the insoluble material and the protein surface.

2.1.1 pH, Surface pK, and Buffering Effects

More recently, the second dimension, that of the microenvironmental effects, has been clearly portrayed. These effects may be summed up as interactions of the substrate solutions with the carrier surface. In this situation, the carrier functional groups act as a buffer or modifying influence upon the bulk solution. The researchers at the Weizmann Institute and The Hebrew University [1,2] demonstrated this effect when they employed highly negatively charged copolymers for immobilizing enzymes. The effect of the carrier upon the pH-activity curve for the immobilized enzyme at low ionic strengths was to shift it to a much more alkaline region than that of the native (soluble) enzyme. This effect could, however, almost be eliminated by increasing the ionic strength with a neutral salt such as sodium chloride. The interpretation offered for these results is that carboxyl groups on the surface of the carrier act as part of the buffering system and participate with the buffer in solution to diminish the pH in the proximity of the enzyme. However, as one increases the ionic strength of a neutral salt, the carrier charge is diminished such that its participation with the buffer in solution is less dramatic.

The view that the Israeli researchers have adopted for their particular results [1] is that the effect on the pH-activity profiles of the bound enzymes can be explained as the result of the effect of the electrostatic potential of the polyelectrolyte carrier on the local concentration of hydrogen ions and positively charged substrate molecules in the micro-environment of the bound enzyme molecules.

Although the interaction of the surface with the substrate molecule certainly does play a most important role, that effect was also noted with an essentially non-charged substrate molecule, glucose. The enzyme, glucose isomerase, was immobilized on a controlled-pore alumina [9] and studied in a continuous column reactor. The substrate was buffered to pH 7.9 with 0.007 M sodium sulfite. After passing through the column of immobilized enzyme, the effluent product was found to be at a significantly lower pH, 7.1–7.7. This pH drop was examined with the carrier that did not contain the enzyme and was found to be an effect of the alumina rather than the composite; thus, there was an actual interaction between the carrier and the feed solutions. A further examination of this study, Fig. 1, reveals that there was an additional shift in the pH-activity curve. The pH levels recorded in parenthesis is that of the effluent. A comparison of these results with those of Takasaki [10] for the soluble enzyme, Fig. 2, will reveal that there was a dramatic shift to the alkaline region with the immobilized enzyme. The optimum pH for the soluble enzyme appears to be narrowly confined to 7.0 while that of the immobilized enzyme appears to be very broad in the region of 7.4–7.7.

Some very general statements may be made with respect to microenvironmental carrier effects upon enzyme reactions. If the surface pK is in the acid region, then the pH-activity curve will be shifted to the alkaline region. Some of these surface effects may be overcome by increasing ionic strength. A further concern is that of the buffer or substrate interaction with the carrier. A substrate that irreversibly reacts with a carrier can essentially lead to a complete loss of enzyme activity as a result of substrate inhibition. This loss of activity may also be noted if a product of an enzyme reaction binds irreversibly to the carrier. The accumulation of the product proximal to the enzyme active site could lead to product inhibition. The surface charge of a carrier may actually repel a substrate of like charge and restrict the reaction of the substrate with the enzyme or it may attract a substrate of opposite charge. The attraction of the substrate may actually enhance the reaction provided that it does not bind irreversibly to the carrier.

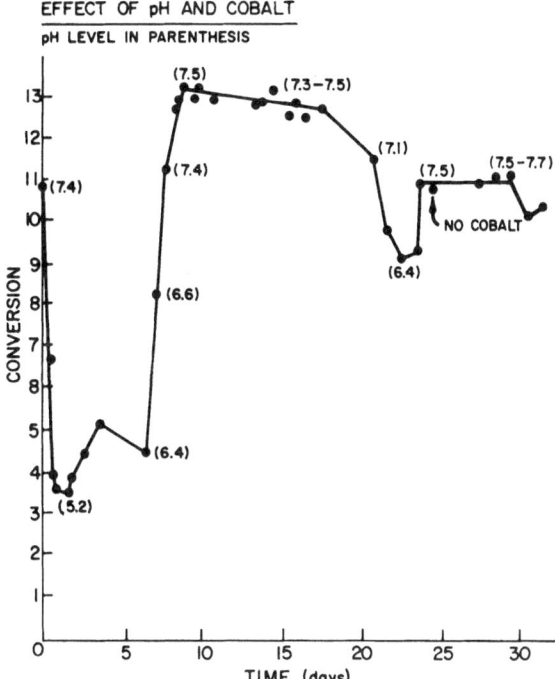

Fig. 1. Effect of pH on
isomerization reaction of
alumina immobilized glucose
isomerase
[from R. A. Messing,
A. M. Filbert, J. Agric.
Food Chem. **23**, 920 (1975)]

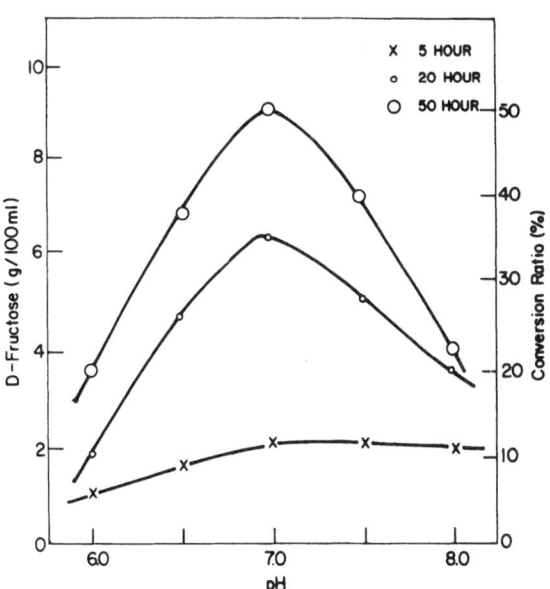

Fig. 2. Effect of pH on isomerization reaction of soluble
glucose isomerase [from
Y. Takasaki, Agric. Biol. Chem.
30, 1247 (1966)]

2.1.2 Hydrophobic and Hydrophilic Carrier Surfaces

In addition to what has been considered the traditional microenvironment effects, the compatibility of both the enzyme and substrate with respect to the environment may play a major role in the effectiveness of the immobilized enzyme. This problem is very simple to visualize if one thinks in terms of lipid enzymes and the problems involved in transporting lipid substrates to the enzyme site in an aqueous environment. If the lipid has a very low solubility in water, then this situation would be very similar to transporting an insoluble substrate to an insoluble enzyme. Now, however, if the enzyme were attached or closely associated with a lipid layer or surface, then the substrate would tend to dissolve preferentially in that layer and gain intimate contact with the enzyme. In fact, many enzyme reactions that occur in living tissues or cells are associated with lipid layers. It is not at all surprising then to find that some of these enzymes have lipid moieties which actually will not bind readily with hydrophilic surfaces and much prefer the hydrophobic environment. These examples are extreme cases. Studies of more subtle effects were performed by Filipusson and Hornby [12] with respect to the enzyme, β-fructofuranosidase, and the inhibition constant of aniline. These studies indicated that the inhibition constant of aniline was about 3 times higher when the enzyme was bound in a hydrophobic carrier.

Johansson and Mosbach [11] performed some rather definitive studies with the enzyme, horse liver alcohol dehydrogenase, by varying the composition of the carrier from hydrophobic to hydrophilic surfaces. A hydrophobic surface was produced with a high ratio of methyl acrylate (hydrophobic monomer) to acrylamide (hydrophilic monomer). In order to produce a hydrophilic surface, the ratio was reversed. These researchers demonstrated that the hydrophobic alcohol, butanol, substrate preferred the enzyme in a hydrophobic matrix while the more hydrophilic substrate, ethanol, did not exhibit that preference.

As a rule of thumb, a carrier having a hydrophilic surface should be chosen for an enzyme that is reacting with a very hydrophilic molecule. A hydrophobic carrier should be chosen when an enzyme reacts with a very hydrophobic substrate. There may, however, be instances where the dictates of the enzyme may be even more critical than those of the substrate. In those instances where an enzyme has a lipid moiety and is far more stable in a hydrophobic environment, the chosen carrier may, in fact, exhibit some hydrophobic characteristics even though it may be reacting with a relatively hydrophilic substrate.

2.1.3 Redox Surfaces

Although oxidation states of carriers have not been generally considered in discussions of the microenvironment, it is the opinion of the author that these exhibit substantial contributions to the effectiveness of many immobilized enzymes. Those enzymes that are readily oxidized appear to be substantially affected by the redox state of the carrier. Typically, the enzymes that are readily oxidized are the sulfhydryl proteases, enzymes containing a great number of sulfhydryl groups on the surface (urease) and enzymes containing attached redox prosthetic groups (glucose oxidase). Glucose oxidase [13], urease [14], and papain, a sulfhydryl protease [15], were studied in the immobilized form on a carrier, titania [16], that exhibits a very powerful reducing surface. The relatively long half-lives of these immobilized enzymes on the reduced surface evidenced the contribution of the carrier.

2.1.4 Metal Ions

Metal ions play an important role in terms of the stability of enzymes and as integral parts of the enzyme-substrate complex. Carriers can, not only, play part in removing these ions from the reaction site and thus reducing either the stability or the activity of the enzyme, but also, they can be employed as contributors of those ions to the enzyme substrate reaction by the proper location upon the surface. The enzyme, glucose isomerase, has two metal ion requirements, magnesium and cobalt. It was demonstrated [9, 17, 18] that these metal ions could be either appropriately fixed or incorporated within the carrier composition such that they contributed to the stability and activity of the immobilized enzyme.

2.2 Side Reactions

Very few, if any, carriers as previously indicated, are inert materials. For this reason, it would not be at all surprising to find that in addition to the immobilized enzyme conversion other (undesired) reactions have been noted with respect to a variety of systems. Quite frequently these side reactions may be attributed to carrier surface contributions.

2.2.1 Non-Specific Adsorption or Accumulations

An immobilized enzyme, independent of the carrier, is charged and thus represents a surface for accumulation. In the author's experience, particularly with enzymes immobilized for sensing in biological fluids, proteins represent the greatest problem in terms of non-specific adsorption. After sufficient serum protein is adsorbed upon the surface, the active site of the enzyme becomes masked and the activity of the preparation is diminished. Large molecules are not the sole offender with respect to this problem. Metal ions may additionally be adsorbed to the surface and may either represent an inhibitor of the enzyme or may be a factor responsible for further adsorption of other high molecular weight materials.

Another problem that the author and others have noted which is, undoubtedly, due to the carrier surface effects is that of accumulations at the surface or entry point to the enzyme. This is not quite the same effect as that of non-specific adsorption although a number of aspects appear to be similar. A specific example of this accumulation is represented in Fig. 3 [15] in which immobilized papain was studied with a casein substrate. Casein, upon acidification or mild hydrolysis, tends to clot or form insoluble particles. In this study, the insoluble particles which may have been caused by either the relatively acid surface of the titania carrier or by the hydrolysis of casein accumulated on the carrier particles such that the mouth of the pore was blocked, thus restricting the flow of the unhydrolyzed casein to the active site of the enzyme. Upon washing, the loosely bound insoluble particles were removed, and the substrate again gained access to the enzyme.

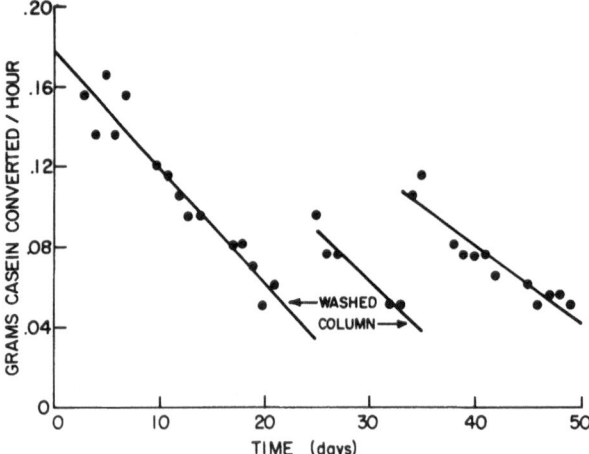

Fig. 3. Performance of Papain immobilized on titania in a continuous plug flow reactor at 37°C with a casein substrate [from R. A. Messing, Immobilized Enzymes for Industrial Reactors, Academic Press, New York (1975)]

2.2.2 Blood Clotting Reaction and Immune Response

There are classes of reactions that are exhibited by carrier surfaces that do not affect the enzyme-substrate reaction; however, these reactions may have an adverse effect on the total system in which the immobilized enzyme is employed. Although the examples that will be cited are medical applications, they are not necessarily limited to this area.

Immobilized enzymes have been employed to interface with blood and tissue either in the form of an extracorporeal shunt or as an insertion in the blood system or tissues for the contribution of an enzyme reaction. These immobilized enzymes have either been used in cases of enzyme deficiencies or for the treatment of a malady such as leukemia. Glass has long been known to be an initiator and an enhancer of the blood clotting reaction. If glass were to be employed as a carrier for an immobilized enzyme that interfaces with the blood, this clotting reaction will ultimately lead to the restriction of the blood flow and the death of the patient.

Another type of carrier that is contraindicated in the case of interfacing with the blood is a carrier that evokes an immune response. Generally speaking, protein carriers with perhaps the exception of pure collagen evoke the immune response. The net result of employing this type of carrier for interfacing with the circulatory system and tissues might be anaphylactic shock followed by death.

2.3 Surface Stability or Durability

One of the most essential considerations in the selection of a carrier for the immobilized enzyme application is that of its surface durability under the application conditions. This is not to imply that a carrier must be durable under all pH, ionic strengths and solvent conditions, but, the surface must be stable under the conditions of attachment and application. If a carrier is undergoing constant erosion or dissolution during the course of the immobilization process, then the enzyme will not be immobilized. On the other hand, if a carrier, such as glass, is employed for an immobilized enzyme that will be used in a very alkaline environment, the half-life of the immobilized enzyme will be extremely short due to the fact that the glass surface will be solubilized under these conditions, and the enzyme will be undercut with its subsequent liberation.

Another problem of concern with respect to a non-durable surface is that of the potential liberation of inhibitors. Even though the backbone of the carrier may be stable and the enzyme may be retained upon the surface, the liberation of an inhibitor from the carrier composition may adversely affect the enzyme substrate reaction.

Eaton [19] has suggested some very realistic static and dynamic evaluation procedures for the durability of carriers under a variety of conditions. Although these procedures were employed primarily for the evaluation of pH conditions, they may just as readily be utilized to evaluate a variety of solvents, buffers, activators, substrates, and products all of which may affect the durability or stability of the surface. The static test that Eaton employed simply involved the immersion of the carrier into a solution of the substances to be evaluated with respect to their effect upon the surface and elevating the temperature to $60°C$ for 16 h. After that period, the samples were washed, and the weight loss of the carrier was determined. Great weight losses indicated that the carriers were not durable under the test conditions.

Once having established a reasonable durability of a carrier in a static process, a more comprehensive evaluation in a dynamic atmosphere should be initiated. Eaton [19] suggested a dynamic durability test which consists of flowing the evaluation solution through the carrier in a downward pattern. This dynamic evaluation may be very realistic for determining the effect upon a carrier in a plug-flow reactor but may, in fact, not be applicable to a continuously stirred reactor or to a fluidized-bed reactor. The evaluation for the durability of a carrier should be performed under the conditions which will closely simulate those of the immobilized enzyme application. Therefore, a dynamic durability test for a continuous stirred reactor should be performed in a continuous stirred mode.

3 Carrier Properties, Morphology, and Configuration

Carriers have frequently been classified according to their composition. Broad classifications such as organics or inorganic materials have been employed to categorize the nature of the carrier. Although there are some merits in these classifications; for instance, organic polymers generally are of lower density than inorganics and thus may be less advantageously employed in a fluidized-bed; these descriptions are not adequate for the full characterization of very pertinent carrier parameters. The morphology of the carrier is extremely important with respect to surface area and pore parameters, both of which in turn will affect the loading of the enzyme.

For the purpose of this discussion, the carriers will be re-classified in the following manner:
1.   Non-porous carriers
2.   Porous carriers
2.1   controlled-pore

2.2   broad pore distribution
2.3   gel structures
2.3.1   preformed gel structures
2.3.2   entrapment gels
2.3.3   copolymers

Superimposed upon this basic morphology is what may be considered the gross configuration of the carrier. Morphology may be considered the microconfiguration while the gross configuration would be the macroconfiguration. The macroconfiguration of a carrier may vary while the microconfiguration remains constant. A carrier can be formed into a particulate, a spherical shape, a fiber, a membrane structure, or a monolith.

3.1 Non-Porous Carrier

A non-porous carrier has one major disadvantage. The surface area of this material is extremely low; therefore, the available surface for the attachment of the enzyme is extremely limited. The enzyme loading problem may be partially overcome by using very fine particles or fibres. However, additional problems arise when materials of this macroconfiguration are employed. Fine fibers or particles are difficult both to remove from the solution and to employ in continuous reactors since they lead to high pressure drops and limited flow rates. On the other hand, non-porous materials such as nylon have been employed advantageously where an interface with the blood system is required. Nylon does not evoke the vigorous clotting reaction that is readily noted with high surface area inorganic carriers such as porous glass. Apparently nylon fibers are compatible with the blood system because of two factors:
1. The basic material,
2. The lower surface area.
Another advantage of utilizing a non-porous carrier is that of the elimination or certainly reduction, of diffusional constraints with respect to the substrate since the enzyme is immobilized on the external surface of the carrier and is in immediate contact with the surrounding environment.

3.2 Porous Carriers

A variety of porous carriers, both organic and inorganic, may be readily obtained in either laboratory or commercial quantities. An example of a controlled-pore organic carrier is the macroreticular polystyrene supplied by Rohm and Haas Company. Controlled-pore inorganic carriers such as controlled-pore glass, controlled-pore silica, and controlled-pore alumina have been produced by Corning Glass Works. The term "controlled-pore" refers to a rather narrow pore distribution. There are many broad pore distribution carriers available which are currently used as catalysts. Porous alumina, a broad pore distribution carrier, has been widely employed as an inorganic catalyst and catalyst support.
The major disadvantage of porous carriers is that most of the surface available for bonding enzymes is internal structure. When an internal structure is employed for bonding the enzyme, not only does the coupling and/or crosslinking reagent require access to the surface, but also, the much larger enzyme and perhaps substrate molecule must un-

dergo these diffusional constraints. An additional problem that arises when a carrier has a broad pore distribution is only a limited number of the pores will be large enough to accomodate both enzyme and substrate; thus, only a small portion of the total surface area will be utilized.

The advantages of employing porous materials are gained from the high surface areas available for the immobilization and from the fact that the enzyme is bonded on an internal surface which is protected from the turbulent and harsh external environment.

3.3 Gel Structure Carriers

A variety of carriers that may be classified as gels have been employed for the immobilization of enzymes. Some of these carriers are collagen [20], starch [21], silica gel [22], acrylamide [23], and cellulose derivatives [24]. These gel structures have been employed for the entrapment or the encapsulation of enzymes and as carriers for the attachment via covalent and/or adsorption coupling. In addition, these structures have been both crosslinked internally and in the presence of the enzyme. These gels do not have truly stable pores and thus it is difficult to measure the size of the pores by mercury porosimetry. The materials are rather elastic in character. The advantages of encapsulation or entrapment may be found in the apparently large loadings that have been achieved without the obvious need of a very stable surface for the attachment. The disadvantage of this type of carrier employed for entrapment or encapsulation procedures are that they are severely limited by diffusional constraints. A very general statement may be made that enzymes immobilized by these procedures may be very useful for small substrates but extremely limited insofar as large molecular weight materials are concerned. When a gel structure is utilized to either couple or adsorb an enzyme, a relatively high surface is available. However, particles of these materials have been found to be rather difficult to utilize in plug-flow reactors since they readily compact and high pressure drops are experienced. Bernath and Vieth [20] and Emery [25] addressed this problem of compaction by winding membranes or sheets containing the immobilized enzymes with alternate layers of backing to form tubular reactors.

3.4 Rigid Versus Elastic Carriers

In order to examine the properties of carriers which are useful for attaching enzymes by either covalent or adsorption procedures, it becomes necessary to reclassify carriers into rigid and elastic (plastic) structures. A few generalizations about the nature of these materials can be made which is partially reminiscent of the old classification of inorganic and organic materials. Inorganic structures are generally rigid while many organic polymers are flexible and elastic. The advantage of the rigid pore structure is that it gives the greatest protection against the harsh and turbulent external environment. In addition, once an enzyme has been immobilized on a rigid surface via many points of attachment, the tertiary structure of the active molecule is maintained by the lack of deformation of the carrier itself.

The main advantage of the elastic carrier is that it may readily be formed into a thin membrane and can be made to conform about another structure. In other words, it can be used as a secondary configuration. In addition to that advantage, generally speaking there are more points by which one can attach an enzyme on an organic elastic surface. A thin membrane would imply a small diffusional path and an optimal available surface for reaction. Another advantage that may be gained from the employment of a membrane is that not only can it be employed for the reaction but also may be utilized for separation. One other point should be made for the employment of elastic carriers particularly in continuous stirred reactors. A rigid structure may readily be abraded in this type of reactor whereas an elastic carrier will be more resistant under these conditions.

It can be readily seen from these points that an advantage gained by one structure leads to a disadvantage in certain applications. Thus, a rigid body is difficult to employ in a thin membrane form, while an elastic carrier may lead to early deformation of the tertiary structure of the enzyme and pressure drops in a plug-flow reactor. To emphasize that point further, a high density carrier, metals and metal oxides, can readily be employed in a fluidized-bed reactor since they are denser than water and may be uniformly distributed, under flow, throughout the reactor while low density carriers (1.0 or under) will accumulate at the upper surfaces of the reactor. Coughlin and Charles [26], demonstrated that metal and metal oxides performed well as carriers in fluidized-bed reactors.

3.5 Microbial Considerations

Another consideration in the selection of a carrier is its resistance to microbial degradation. Carriers which are rich in carbon, such as starch, or rich in carbon and nitrogen, such as protein, are potentially good nutrients for microbes. If the carrier is attacked by microbes, the enzyme is then released into solution. Generally speaking, inorganic materials such as inorganic oxides (silica, titania, alumina etc.) are resistant to microbial attack. However, there are also organic materials, i.e. fluorocarbons, polypropylene, etc. which are resistant to microbes. One should also appreciate the fact that microbes will attack and destroy enzymes since they are proteins. If an enzyme, however, is buried in a pore having a diameter of less than 1000 Å, it would be impossible for microbes which generally exhibit dimensions greater than 1000 Å to gain access to this enzyme. Even though the enzyme would be excluded from the microbial environment, one must bear in mind that some microbes elaborate exocellular proteases that may gain access to and destroy the immobilized enzyme within the pore.

It should be borne in mind that a material does not necessarily have to be totally resistant to microbial attack to be useful as a carrier. Vieth and Venkatasubramanian [27] state that collagen is resistant to proteolytic attack and with appropriate tanning treatment they have not observed microbial degradation of the matrix under actual conditions of use and storage. Cellulose, also, represents a carrier that is not very rapidly attacked by microbes. Other synthetic, organic carriers have been found to be resistant to microbial attack. If the feed (substrate) solutions are delivered to the reactor substantially free of microbes, there may actually be no concern with respect to microbial degradation provided the reactor containing the immobilized enzyme is free of microbes.

3.6 Thermal Stability of Carrier

The thermal stability of a carrier plays a role in both the bonding and utilization of the enzyme. Although enzymes are normally exposed to a very limited temperature range, $0-80\,°C$, a carrier having a large coefficient of expansion within this range may present numerous problems. For example, if an enzyme is attached to a carrier that has a large coefficient of expansion, the enzyme active site may be distorted or destroyed as the carrier expands or contracts with changing temperature. This could occur during the immobilization of the enzyme where several different temperatures are required in a multi-step immobilization, or during the processing of the substrate in the continuous reactor where the reaction is brought from room temperature to operational temperature and may involve an elevation of as much as $45\,°C$. Relative to cellulose and other organic carriers, the inorganics have low thermal coefficients of expansion and are dimensionally stable over this range.

3.7 Carrier Shape and Particle Size

The shape or particle size of the carrier can affect the performance of the immobilized enzyme in a continuous reactor. The larger the particle, the less the pressure drop; therefore, a uniform high flow rate can be achieved with large particles. However, counterbalancing this, is the negative aspect that with large particles, a greater diffusion path must be negotiated by the substrate to reach all of the active enzyme sites. As a matter of fact, it has been noted that with very large particles only the enzyme molecules immobilized in the outer $5-10\%$ of the particles are fully utilized. Of course, where the substrate solution is either extremely viscous or contains particles it may not be wise to choose a plug-flow reactor. Under these circumstances, a fluidized-bed or a fixed carrier configuration such as a monolith may be more judicious choices.

3.8 Preliminary Carrier Selection

Tab. 1 entitled "Properties Suggesting Carrier Applications" contains information for the preliminary selection of appropriate carriers based upon the specific property of the carrier and matching it to the application condition. This is not to indicate that these are the carriers which would be chosen under all circumstances for the specific application; however, the table is a fairly useful guide to a rational first choice.

3.9 Optimizing a Dimensionally Stable Controlled-Pore Carrier

One of the most important considerations for optimizing an appropriate dimensionally stable carrier is that of the pore diameter. The pore diameter, or more specifically the bulk of the pore diameters within a carrier, should be chosen such that it is based upon the major axis of the enzyme unit cell, if, the enzyme is larger than the substrate. On the other hand, if the major dimension of the substrate is greater than the major axis of the enzyme unit cell, then the pore dimension of the carrier should be chosen with respect to the substrate.

The term "pore diameter" was used several times in the previous paragraph. It is necessary to define that terminology with respect to the procedure for its determination. The

Table 1. Properties Suggesting Carrier Applications

Carrier type	General application	Conditions suggesting application	Reference
Silastic (silicone rubber)	Medical interface w/ blood or tissues	Do not initiate clotting reaction, little or no immune response evoked	[28]
Nylon	Same	Same	[29]
Collodion	Same	Same	[30]
Methacrylate	Same	Same	[31]
Collagen	Polymer hydrolysis and collection of products free of substrate	Membranes to separate substrate and products (protein hydrolysis), continuous stirred reactor	[20,27]
Collodion	Same	Same	[32]
Controlled-pore glass or silica	Continuous plug-flow and fluidized-bed reactors at pH's below 7.0	Dimensional stability, low pressure drops, durable below pH 7.0	[33–35,19]
Controlled-pore alumina	Continuous plug-flow and fluidized-bed reactors, pH 5–11	Dimensional stability, low pressure drops, durable above pH 5.0	[34,35,19]
Controlled-pore titania	Continuous plug-flow and fluidized-bed reactors, pH 3–9	Dimensional stability, low pressure drops, durable pH 3–9	[34,35,19]

method most commonly used to determine pore diameters of inorganic materials is that of mercury intrusion porosimetry. The most widely used equipment for this determination is the Aminco-Winslow Porosimeter. The determination is performed by forcing mercury into the pores of the carrier under pressure. The amount of penetration in cc/gm is measured with increasing pressure in terms of pounds/inch^2. The pressure required for the mercury to penetrate the pores is correlated to the diameter of the pore. These data are displayed in Fig. 4. The more vertical the line representing the penetration, the closer the pore distribution is throughout the carrier. In other words, a vertical line represents a tightly controlled pore carrier while a gently rising slope indicates a very broad pore distribution within the carrier. This procedure, in fact, measures the size of the bottlenecks or the diameter of the narrowest passages which restrict the flow of mercury. The widest chambers in the form of a bubble or wide passageway beyond the narrow restriction are not measured. A pore, then, is actually a restriction of access to internal surfaces and volumes.

It became rather apparent in some of the early work with controlled-pore glass [36] that the amount of enzyme adsorbed by the carrier was not only a function of the surface area but also of the pore diameter. Even though a carrier material may have a high surface area, the availability of some of that surface for attachment is limited by the restrictive openings. An example of a material that has an extremely high surface area but little available internal surface for attachment is bentonite. Another example of such a material is controlled-pore glass with an average pore diameter of substantially less than 100 Å.

A rather firm comprehension of the relationship between pore diameter and molecular dimension was evidenced in studies with controlled-pore ceramics [13]. The objective

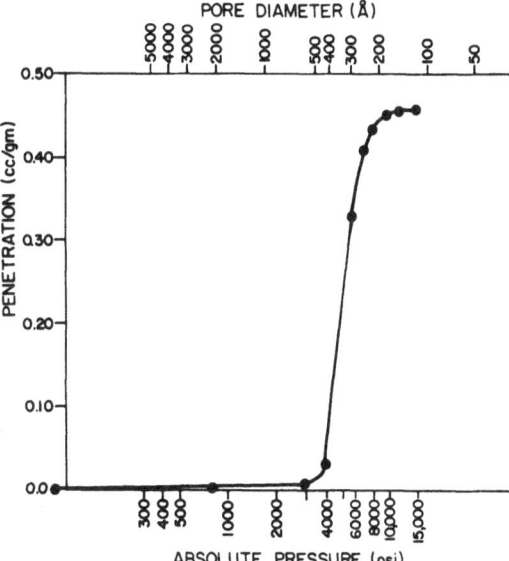

Fig. 4. Pore determination by mecury intrusion porosimetry [from R. A. Messing, Immobilized Enzymes for Industrial Reactors, Academic Press, New York (1975)]

of this program was to fully utilize the internal surface of porous carriers for coupling the enzyme and yet maintain as high a surface area as possible. To attain this objective, the smallest pore diameter with a very narrow pore distribution is required to just allow the entrance of the limiting molecule (the enzyme when it is larger than the substrate). Studies of glucose oxidase and catalase with relation to pore diameter indicated that the optimum pore diameter for immobilizing an enzyme was approximately two times the major axis of the unit cell of the enzyme.

In this study, it was demonstrated that catalase protects glucose oxidase from the destructive effects of peroxide. Hydrogen peroxide, one of the products of glucose-glucose oxidase reaction, accumulates within the pores. In the absence of catalase, glucose oxidase is destroyed by the peroxide. When catalase is present within the pore, this enzyme loses no activity over a two month period, and losses of only 18% of the original activity were noted with storage at room temperature in water over a 5 1/2 month period.

The largest dimension of the glucose oxidase unit cell is 84 Å. This enzyme should have substantial access through a pore diameter having 168 Å, twice the major axis. Catalase, on the other hand, having a major axis of 183 Å, would not appear to be restricted by a pore diameter approximating 366 Å. The studies that were performed indicated that glucose oxidase entered alumina pores of 175 Å, however, this immobilized enzyme was exceedingly unstable due to the exclusion of catalase. On the other hand, a very stable glucose oxidase-catalase system was achieved with a titania having an average pore diameter of 350 Å and a maximum pore diameter of 400 Å. The highest loading, Fig. 5,

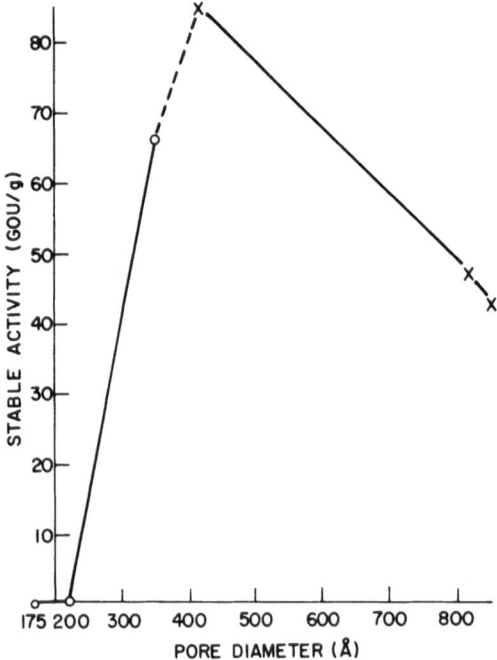

Fig. 5. Cumulative data relating
stable glucose oxidase activity to
pore diameter [from R. A. Messing,
Biotechnol. Bioeng. **16**, 897 (1974)]

of stable glucose oxidase-catalase was achieved with a porous titania having an average
pore size of 420 Å with a minimum pore diameter of 300 Å and a maximum pore diameter of 590 Å. The surface area and pore relationship of the carriers employed for
this study are recorded in Tab. 2. It may be noted that if one were to take the data,
plotted in Fig. 5, disregard the dashed line and extrapolate the two solid lines such that

Table 2. Carriers Employed in Glucose Oxidase Pore Diameter Studies

	Al_2O_3	44% Al_2O_3 56% Al_2O_3	TiO_2	TiO_2	TiO_2	TiO_2
Average pore diameter (Å)	175	220	350	420	820	855
Minimum pore diameter (Å)	140	140	220	300	760	725
Maximum pore diameter (Å)	220	300	400	590	875	985
Pore volume (cm^3/g)	0.6	0.5	0.45	0.4	0.2	0.22
Surface area (m^2/g)	100	75	48	35	7	9
Particle mesh size	25/60	25/60	25/60	30/80	25/80	25/80

they intersect, the intersection would approximate 370 Å for the optimum loading. It, therefore, becomes apparent that twice the major axis or spin diameter of the largest molecule in the system, catalase, is the limiting factor. Catalase has a spin diameter of 366 Å and that is very close to the extrapolated value of 370 Å.

4 Modification and Preconditioning of Carriers Prior to Immobilization

Carriers for the most part are very active materials. A prime concern should be that these materials are clean and fully activated with respect to their functional groups prior to the immobilization process. A carrier such as controlled-pore glass will adsorb volatile substances from the air as well as microbial films. If a piece of porous glass is exposed to room atmosphere within three to four weeks it will turn from a clear-white color to an amber-brown. These are organic substances that are contributed by the atmosphere and will, of course, mask the functional silanol groups on the surface. If this dirty glass were to be employed for an immobilization process either by adsorption or covalent coupling, the loadings of the enzyme or coupling reagents would be extremely low and perhaps loosely bonded. A variety of acid and base cleaning techniques along with sonication have been employed to clean controlled-pore ceramics; however, the simplest procedure involves simply exposing the carrier to temperatures in excess of 450°C in the presence of sufficient air or oxygen to volatize the carbonacious material. For relatively small quantities of carrier, a simple furnace elevated to 500°C for 1 1/2 h without an additional supply of air is adequate to thoroughly clean the internal as well as the external surfaces.

The cleaning processes for most carriers will be dependant upon the basic durability characteristics of the carrier. For instance, a carrier that is very stable in an acid environment should be cleaned with acid. On the other hand, a carrier that is stable in the basic region should be cleaned with base. Generally organic carriers cannot be treated with a pyrolysis process for cleaning surfaces. However, a variety of organic solvents such as alcohol, acetone, carbon tetrachloride may be chosen for removal of foreign films. Again the durability of the carrier in these solvents should be ascertained before the treatment.

A problem that is frequently encountered concerns cross-linking of the enzyme as a side issue to covalent coupling on a derivatized carrier. The crosslinking occurs as a result of the failure to remove excess coupling agent prior to the coupling procedure. Generally adequate washing of the carrier with the solvent that is employed for the application of the coupling agent to the surface will overcome this problem. In many cases, another solvent may be employed to thoroughly remove the coupling agent but the choice of the additional solvent should be such that the functional group is not destroyed in its presence.

Perhaps the simplest but the most frequently forgotten process is that of preconditioning the carrier for the immobilization of the enzyme. Each enzyme has its own stability characteristics with respect to pH, ionic strength, activators, and cofactors. A common error, quite frequently encountered, is that an enzyme solution is applied directly to the surface of a dry carrier without prior exposure to the preferred pH, salt, etc.

conditions. The result, of course, is a dramatic loss in total activity. These losses may be attributed either to the very adverse pH environment contributed by the non-pre-conditioned surface or the removal of the stabilizing salt from the enzyme environment by the carrier. Prior to the immobilization process, the carrier should be treated with sufficient volumes (at least 10 volumes of solution for each volume of carrier) for sufficient lengths of time (greater than one hour at room temperature) with the buffer- activator solution which offers the greatest stability to the particular enzyme that will be immobilized.

Most carriers selectively bind such cofactors as metal ions. The carrier, therefore, is in competition with the enzyme requiring metal ions for activation. Hence, these metal ions will be removed from the functional site of the enzyme and bound to the carrier. The enzyme will then be reduced in activity or become inactive. We have found that in a number of instances it is appropriate to precondition the surface of the carrier prior to exposure of the enzyme with a solution of metal ion that is required for the enzyme. Pre-treatment of the carrier with a concentration of metal ions approximately ten times higher than that required for activation of the enzyme in a volume at least twice the volume of the carrier for approximately 1 h at $37\,^{\circ}$C has generally been effective for saturating the carrier requirements for the metal ion. This pretreatment will generally avoid the competition by the carrier for the enzyme metal ions.

5 Regeneration of Systems

There are three separate entities to be considered under this topic:
1. Reactivation of the immobilized enzyme
2. Regeneration of the immobilized enzyme
3. Regeneration of the carrier.

5.1 Reactivation of the Immobilized Enzyme

In this case, the immobilized enzyme is normally reversibly inactivated. This may be due to a plugging of the pores in a controlled-pore ceramic, Fig. 3, or a film formation on the surface of the immobilized enzyme. Generally this can be handled by simple washing procedures for the immobilized preparation.

5.2 Regeneration of Immobilized Enzyme

In this case, the enzyme has irreversibly lost activity. Such was the case with aminoacy-lase immobilized on DEAE-Sephadex reported by Chibata et al. [37]. These authors found that if they exposed the somewhat inactivated immobilized enzyme to fresh enzyme solution, they could regenerate the full activity of the original immobilized enzyme preparation.

5.3 Regeneration of the Carrier

When an immobilized enzyme loses activity and cannot be regenerated simply by exposing the preparation to fresh enzyme, and the useful level of activity for the preparation has passed, then, it may be wise to remove the inactive enzyme and prepare the carrier surface for a fresh immobilization. This carrier regeneration can be accomplished by either changing the pH such that the enzyme is desorbed from the carrier or the enzyme can be chemically stripped from the surface by such reagents as hydrogen peroxide and other oxidizing agents. As suggested in the previous section, the simple pyrolysis procedure may be employed for the regeneration of inorganic carriers.

6 Methods for Immobilizing by Adsorption

The purpose of this section is to identify problems concerning the carrier during the immobilization process. The immobilization process of adsorption was chosen as a representative process.

There are four distinct methods for immobilizing an enzyme by adsorption:
1. The static procedure
2. Electrodeposition
3. Reactor loading process
4. Mixing or shaking bath loading

6.1 The Static Procedure

This process is perhaps the most inefficient of the four methods and requires the most time. The carrier is loaded by simply allowing the solution containing the enzyme to contact the carrier without agitation or stirring. The enzyme is diffused to the carrier surface and finally into the internal structure. Generally the loading is not uniform and is rather low unless the carrier is exposed to the enzyme for many days. There is no particular precaution required in the handling of the carrier in this process.

6.2 Electrodeposition

Electrodeposition is an interesting process whereby the carrier is placed proximal to one of the electrodes in an enzyme bath, the current is turned on, the enzyme migrates to the carrier and is deposited upon the surface. There are two concerns with respect to the carrier when this process is employed. The first concern is the durability of the carrier under the electrical field. This should be analyzed prior to the immobilization. The second concern reflects removal of the ions that may have been placed upon the surface in the preconditioning treatment of the carrier. The removal of those ions may lead to high activity losses during the process.

6.3 Reactor Loading Process

This procedure is perhaps the prime candidate for commercial preparation of immobilized enzymes. The carrier is loaded into the reactor that will be employed for the processing system. The enzyme is then added to the reactor and the carrier is loaded in a dynamic environment which may be achieved either through circulation or agitation. This approach may be employed with continuous stirred reactors, plug-flow reactors, or fluidized-bed reactors. The specific concerns with respect to the carrier are those of durability or pressure drop reflected in achieving the dynamic environment necessary for loading the carrier. This is somewhat reminiscent of the discussions concerning morphology and configuration in Section 3. In other words, if the final reactor process is to be plug-flow and the carrier is a rigid structure, then agitation should not be the method of choice for loading the enzyme in the reactor. On the other hand, if the final processing of the feed is to be in the continuous stirred reactor mode and an elastic carrier is employed, one should not attempt to circulate the enzyme in a plug-flow manner such that high pressure drops are experienced. The recommendation is that the enzyme loading process should employ a similar mode as will be employed for the final feed-to-product process.

6.4 Mixing or Shaking Bath Loading

This technique is probably employed most frequently for laboratory preparations. The carrier is placed in the enzyme solution and either mixed with a stirrer or placed in a shaking water bath and continually agitated. The process is effective, and normally results in rather uniform loading. The particular concerns with respect to the carrier are:
1. Too vigorous stirring may destroy a rigid structure,
2. A low density carrier may float to the surface and not be adequately exposed to the enzyme solution.

7 Approaches to Compensation for Surface Contributions

There are two objectives in modifying the surface of a carrier which is to be used for the immobilization of enzymes. The first purpose is termed "activation". The intent in this case is to derivatize or functionalize the surface generally for the covalent attachment of enzymes. This discussion will not be versed toward derivatization of surfaces. The second goal that is desired when a surface is modified is that of changing some surface contribution or effect that was discussed under topic 2, "Surface Contributions of the Carrier".
It should be borne in mind that surfaces of carriers may be readily modified to offer desired properties; however, in the course of the application the effect of the modification may be reduced or undercut such that the carrier will revert to its original condition. Ideally speaking, if a carrier could be found that has a uniform composition and fulfills all the requirements necessary, it is preferable to use that carrier rather than attempting to modify a carrier surface for a specific application. There are many instances, of course, when the properties of two very different carriers, such as that of a rigid carrier base but a very elastic surface, are desired and certainly under those conditions, the modification of the carrier surface is the most sensible approach.

7.1 Zirconia Surface Treatment of Glass for Durability

Under an alkaline environment, glass surfaces are rapidly eroded. In addition, glasses that display high surface areas tend to become very hydrophilic. Under these conditions, the glass surface tends to react readily with a variety of reagents and thus, durability problems may ensue. Tomb and Weetall addressed this problem by applying a coating of zirconia [38, 39] to the glass surfaces. Weetall *et al.* [40, 41] demonstrated that this approach increased the durability of glass under application conditions which were not necessarily in alkaline pH environment. It should be pointed out that zirconia coating increases the alkaline durability of glass; however, under extended conditions of high pH, considerable quantities of silica may be lost from the carrier.

7.2 Stannous Treatment of Titania Surfaces

An approach to converting the surface of titania by employing stannous salts for surface treatment [14] indicated the very minimum that a more compatible situation for the enzyme urease could be achieved. Actually, in this particular case, it would appear as though the surface was modified with respect to its redox potential.

7.3 Aromatic Surfaces to Increase Durability

An additional approach to increasing the durability of a carrier in an aqueous environment is to modify the surface such that it repels water. To accomplish this end, biphenyl structures were attached directly to silica surfaces [42, 43]. This hydrophobic treatment of the surface did increase not only the aqueous durability of the carrier but also allowed the carrier to operate in a slightly alkaline environment, pH 7.8, [42] under which conditions this carrier, generally, is not recommended. By employing this approach, it was found that the half-life of the immobilized enzyme could be increased by approximately 1.8 times.

8 Concluding Remarks and Additional Precautions that Concern the Carrier and the Immobilized Systems

Immobilized enzymes are not the panacea for all enzyme applications that was initially projected in the late 1960's and the early 1970's. As a matter of fact, at this point in time we are able to focus our attention on those applications which may derive an economic advantage in the application of the immobilization technology. Recent evidence confirms the fact that we do not change the basic operating parameters of the enzyme by attaching it to a carrier but rather the advantages are found in the multiple use and the control of the microenvironment.

In the early studies of immobilized enzymes, half-lives were projected of greater than one year on studies that involved two or three months duration. This approach was, of course, proved to be fraught with erroneous conclusions. In addition, the carrier which was frequently termed inert, was minimized with respect to its contribution. Furthermore, coupling techniques were not adequately examined with respect to their gross and specific effects upon the enzyme. All of these factors appear to be scrutinized rather closely in the current endeavors. It is with this view in mind that the efforts in immobilization should be considered as a multi-disciplinary study for the harmonizing of the enzyme with the carrier, coupling technique and the application environment.

This review has addressed the immobilization of enzymes, however, the technology of immobilization may have its larger applications in other disciplines. These disciplines include fermentation, immunoassay, and affinity chromotography. It should be noted that the characteristics of the carrier in these applications must also be matched to the environment of the application. There will, undoubtedly, be differences such as the morphology of the carrier with respect to surface area and pore parameters.

9 References

1. Goldstein, L., Levin, Y., Katchalski, E.: Biochemistry 3, 1913 (1964)
2. Levin, Y., Pecht, M., Golstein, L., Katchalski, E.: Biochemistry 3, 1905 (1964)
3. McLaren, A. D., Estermann: Arch. Biochem. Biophys. 68, 157 (1957)
4. Messing, R. A.: In: Volume II "Glass '77". J. Gotz (ed.). CVTS-DUM Techniky, Publ. Prague, Czech., 1977 p. 484
5. Abramson, H.A., Moyer, L.S., Gorin, M.H.: Electrophoresis of Proteins: Reinhold Publishing Corp., New York, 1942
6. Abramson, H.A.: J. Gen. Physiol. 15, 575 (1932)
7. Moyer, L.S.: J. Phys. Chem. 42, 71 (1938)
8. Tiselius, A.: Nova Acta soc. sc. Upsaliensis 7, (No.4) (1930)
9. Messing, R.A., Filbert, A.M.: J. Agric. Food Chem. 23 (1975)
10. Takasaki, Y.: Agric. Biol. Chem. 30, 1247 (1966)
11. Johansson, A.C., Mosbach, K.: Biochim. Biophys. Acta 370, 339, 348 (1974)
12. Filipusson, H., Hornby, W. E.: Biochem. J. 120, 215 (1970)
13. Messing, R. A.: Biotechnol. Bioeng. 16, 897 (1974)
14. Messing, R. A.: Biotechnol. Bioeng. 16, 1419 (1974)
15. Messing, R. A.: "Immobilized Enzymes for Industrial Reactors." Messing, R. A. (ed.). Academic Press, New York, 1975 p. 81
16. Messing, R. A.: Process Biochemistry 9, 26 (1974)
17. Eaton, D. L., Messing, R. A.: U.S. Patent 3,982,997 (9/28/76)
18. Eaton, D. L., Messing, R. A.: U.S. Patent 3,992,329 (11/16/76)
19. Eaton, D. L.: In: "Immobilized Biochemicals and Affinity Chromotography". Dunlap, R. B. (ed.) Plenum Publishing Corp., New York, 1974
20. Bernath, F. R., Vieth, W. R.: In: Immobilized Enzymes and Microbial Processes. Olson, A. C., Cooney, C. L. (eds.). Plenum Press, New York, 1974
21. Bauman, E. K., Goodson, L. H., Guilbault, G. G., Kramer, D. N.: Anal. Chem. 37, 1378 (1965)
22. Johnson, P., Whateley, T. L.: J. Colloid Interface Sci. 37, 557 (1971)
23. Updike, S. J., Hicks, G. P.: Nature 214, 986 (1967)
24. Dinelli, D., Marconi, W., Morisi, F.: In: Methods in Enzymology, Mosbach, K. (ed.), Vol. XLIV, Academic Press, New York, 1976 p. 227
25. Emery, A.H.: In: Enzyme Engineering 2, Pye, E. K., Wingard, L. B., Jr. (eds.). Plenum Press, New York, 1974 p. 269
26. Coughlin, R. W., Charles, M.: In: Enzyme Engineering 2, Pye, E. K., Wingard, L. B., Jr. (eds.). Plenum Press, New York, 1974 p. 339
27. Vieth, W. R., Venkatasubramanian. In: Methods in Enzymology. Mosbach, K. (ed.). Vol. XLIV, Academic Press, New York, 1976 pp 244, 258
28. Pennington, S. N., Brown, H. D., Patel, A. B., Chattopadhyay, S. K.: J. Biomed. Mater. Res. 2, 443 (1968)
29. Chang, T. M. S.: Biochem. Biophys. Res. Commun. 44, 1531 (1971)
30. Chang, T. M. S. Artificial Cells. Thomas, Springfield, IL (1972)
31. Hersh, L. S.: Trans. Amer. Soc. Artif. Int. Organs 18, 54 (1972)
32. Goldman, R., Silman, H. I., Caplan, S. R., Kedem, O., Katchalski, E.: Science 150, 758 (1965)

33. Messing, R. A.: J. Non-Crystalline Solids 19, 277 (1975)
34. Messing, R. A.: Immobilized Enzymes for Industrial Reactors, Messing, R. A. (ed.). Academic Press, New York, 1975
35. Messing, R. A.: Research/Development 25, 32 (1974)
36. Messing, R. A.: Enzymologia 39, 13 (1970)
37. Chibita, I., Tosa, T., Sato, T., Mori, T., Matsuo: In: Fermentation Technology Today. Terui, G. (ed.). Society of Fermentation Technology, Yamada-Kami, Suita-shi, Osaka, Japan, 1972 p. 383
38. Tomb, W. H., Weetall, H. H.: U.S. Patent 3,783,101 (1/1/74)
39. Tomb, W. H., Weetall, H. H.: U.S. Patent 4,025,667 (5/24/77)
40. Weetall, H.H., Havewala, N.B.: Biotechnol. Bioeng. Symp. 3, 241 (1972)
41. Weetall, H. H., Havewala, N B., Garfinkel, H. M., Buehl, W. M., Baum, G.: Biotechnol. Bioeng. 16, 169 (1974)
42. Messing, R. A., Stinson, H. R.: Molec. Cell Biochem 4, 217 (1974)
43. Messing, R. A., Bialousz, L. F., Lindner, R. E.: J Solid Phase Biochem 1, 151 (1976)

Industrial Applications of Immobilized Biocatalysts

P. Brodelius*
Department of Chemistry, Q-058
University of California at San Diego, CA 92093, USA

Contents

1	Introduction	76
2	Engineering Aspects	77
	2.1 Sources and Isolation of Enzymes	77
	2.2 Preparation of Immobilized Enzymes	78
	2.2.1 Adsorption	78
	2.2.2 Covalent Linkage	79
	2.2.3 Entrapment and Encapsulation	80
	2.2.4 Cross-Linking	81
	2.3 Characteristics of Immobilized Enzymes	81
	2.3.1 Stability of Immobilized Enzymes	81
	2.3.2 Kinetic Behavior	82
	2.3.3 Multienzyme Systems	83
	2.3.4 Whole Cell Systems	83
	2.4 Reactor Design and Performance	84
	2.4.1 Batch Reactors	84
	2.4.2 Continuous-flow Reactors	84
	2.4.3 Choice of Reactor Type	85
3	Application in Food Industry	86
	3.1 Sugar Industry	86
	3.1.1 Hydrolysis of Polysaccharides	86
	3.1.2 Isomerization of Glucose	89
	3.1.3 Hydrolysis of Sucrose	91
	3.1.4 Hydrolysis of Raffinose	92
	3.1.5 Production of Gluconic Acid	93
	3.2 Dairy Industry	94
	3.2.1 Hydrolysis of Lactose	94
	3.2.2 Cheese Manufacturing	96
	3.2.3 Sterilization of Dairy Products	97
	3.2.4 Miscellaneous Applications	97
	3.3 Production of Amino Acids	99
	3.3.1 Resolution of Racemic Mixtures	99
	3.3.2 Biosynthetic Production of Amino Acids	102
	3.3.3 Hydrolysis of Proteins	106
	3.4 Miscellaneous Applications in Food Industry	107
	3.4.1 Beer Industry	107
	3.4.2 Juice and Wine Industries	108
	3.4.3 Production of Nucleotides	109

* Present address: Biochemical Division, Chemical Centre,
University of Lund, S-22007 Lund 7, Sweden

4   Applications in the Pharmaceutical Industry 110
    4.1 Penicillins 110
    4.2 Cephalosporins 113
    4.3 Steroids 116
    4.4 Organic Acids 117
         4.4.1 α-Keto-Acids 117
         4.4.2 Urocanic Acid 118
         4.4.3 L-Malic Acid 118
         4.4.4 2-Keto-L-Gulonic Acid 119
    4.5 Fine Chemicals 119
         4.5.1 Coenzyme A 120
         4.5.2 Glutathione 120
         4.5.3 Porphobilinogen 120
    4.6 Radioactive Compounds 121
5   Applications in Waste Treatment 122
6   Concluding Remarks 123
7   References 124

The isomerization of glucose to fructose, the production of amino acids, and the hydrolysis of penicillins to 6-aminopenicillanic acid are some examples of the industrial processes, based on immobilized biocatalysts, that are operative at present. These and other processes utilized in the food and the pharmaceutical industries, as well as a number of potentially useful immobilized biocatalysts, which have been investigated on pilot plant or laboratory scale, are discussed in this review. A general development of the engineering aspects of the methods will precede the description of specific applications of immobilized enzymes and microbial cells.

1 Introduction

The utilization of enzymes for the manufacturing of products such as cheese, bread and alcoholic beverages dates back many centuries. During more recent years a number of processes involving enzymes as catalysts have been developed. The advantages of enzymes over chemical catalysts are many and well established. The use of enzymes in industrial applications has been so far limited to certain food and drug products by several factors. Many of these limitations can be overcome by using immobilized enzyme technology and during the last decade enzyme engineers have shown an increasing interest in this technology. A look at some of the advantages of immobilized enzymes over soluble enzymes can explain this growing interest. Expensive enzymes, when immobilized, can be re-used. Processes involving immobilized enzymes can be operated continuously and readily controlled. Furthermore products are easily separated from the enzyme catalysts and in some cases enzyme properties (e.g. stability) can be altered favorably by immobilization.

To date, only a few processes based on immobilized enzymes are operative on an industrial scale. All of these applications are examples of relatively simple enzyme reactions such as hydrolysis and isomerization. More complex reactions involving coenzymes have not yet been developed for industrial use to any extent. Such reactions, however, have a great potential for utilization in industrial catalysis. In this review immobilized enzyme technology currently used in industry, as well as a number of potentially useful commercial applications of these techniques, will be discussed.

A general development of the engineering aspects of the methods will precede further elucidation of specific applications of immobilized enzymes on an industrial scale.

2 Engineering Aspects

2.1 Sources and Isolation of Enzymes

Enzymes used in industry are isolated from animal and plant tissues, as well as microorganisms. During recent years microorganisms have become increasingly important as producers of industrial enzymes and in fact most enzymes used in industry today are of microbial origin.

Attempts are now being made to replace enzymes which traditionally have been isolated from animal tissues (i.e. trypsin, chymotrypsin, pepsin, rennin) and plants (i.e. papain, bromelain, ficin) with enzymes from microorganisms. An explanation for such a trend is the fact that the production of enzymes from animal glands, for example, is highly dependent on the production of and demand for meat. On the other hand, a constant supply of enzymes from microorganisms can be easily obtained. A recent example of such an effort is rennet, which is used to curdle milk in cheese manufacturing. Rennet from the fungi *Mucor mietrei* and *Mucor pusillus* has largely replaced a similar enzyme isolated from calf stomachs during the last few years [1].

Another factor which has influenced and will further promote the increasing use of microbial enzymes is the possibility of genetically manipulating microorganisms to produce abnormal amounts of enzymes. This genetic engineering can be achieved by recombination or more conventional methods such as induced mutations and changes of the environment in which the microbe is grown. Such enhanced production of a particular enzyme can simplify the extent of purification and isolation that may be needed, as well as reduce the capital investment in fermentation plants.

The utilization of thermophilic microorganisms for the production of heat stable enzymes is another example of recent progress in enzyme engineering. Heat stable enzymes are particularly attractive for use in enzyme reactors. Processes can be operated at higher temperatures resulting in faster reaction rates, decreased fluid viscosity and increased reactant and product solubility. In addition to this, the risk of detrimental effects of microbial contamination in the reactor is reduced.

A variety of enzyme purification procedures are currently available. A summary of the steps involved in the commercial production and purification of enzymes is shown in Fig. 1. As can be seen, a number of steps are required for the final standardized product. The actual purification, which takes place in the fractionation step, often involves several operations such as chromatography (ion-exchange and/or gel filtration), electrophoresis and fractional precipitation. As we shall see later, enzymes do not necessarily need to be highly purified for use in an immobilized state. In fact, very crude preparations, even whole microbial cells, have been used successfully on an industrial scale. The required degree of purification depends on several factors such as possible side reactions, enzyme stability and cost.

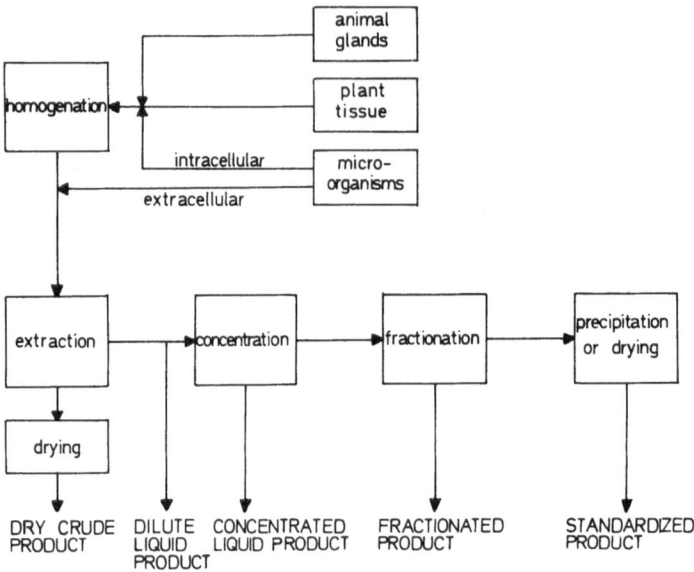

Fig. 1. Industrial production and purification of enzymes

2.2 Preparation of Immobilized Enzymes

There have been a large number of methods reported for immobilizing enzymes. Most of these methods can be classified under one of the following categories or may represent combinations thereof:
1. adsorption
2. covalent linkage
3. entrapment or encapsulation
4. cross-linking

Tab. 1 summarizes some of the relative advantages and disadvantages of these different types of immobilization techniques. No one technique will serve well in all cases, because enzymes differ widely in their composition and overall chemical characteristics. In addition to this, substrate and product properties will influence the choice of immobilization technique.

Next, a brief presentation of the most commonly used techniques and supports will be given. For more detailed information a number of books and reviews are available [2—8].

2.2.1 Adsorption

An adsorbed enzyme is retained on a solid support by forces other than covalent linkages (e.g. ionic, hydrogen or hydrophobic bonding). The binding strength between enzyme and matrix is in most cases relatively weak and therefore often sensitive to chan-

Table 1. Comparison of different immobilization techniques

Characteristics	Adsorption	Covalent linkage	Entrapment and encapsulation	Cross-linking
Preparation	Simple	Difficult	Difficult	Intermediate
Binding force	Weak	Strong	Intermediate	Strong
Enzyme activity	Intermediate	High	Low[a]	Low
Regeneration of the carrier	Possible	Rare[b]	Impossible	Impossible
Cost of immobilization	Low	High	Intermediate	Intermediate
General applicability	Yes	No	Yes	No
Protection of enzyme from microbial or proteolytic attack	No	No	Yes	Somewhat

[a] Low specific activities often due to diffusion barriers.
[b] Some carriers can be regenerated (e.g. porous glass).

ges in the environment, such as pH, ionic strength, and/or temperature. This reversibility of attachment requires a tight control of process conditions to avoid leakage of enzyme. A great advantage of this method, however, is the simplicity in preparing the immobilized enzyme. In most cases, immobilization is achieved simply by bringing the enzyme solution in contact with the adsorbent surface. A suitable adsorbent should possess high affinity for the enzyme and yet cause minimal denaturation. It should also have a high protein binding capacity. Enzymes have been adsorbed on many different supports such as organic polymers, mineral salts, metal oxides, and various siliceous materials. As will be discussed later ion-exchangers (e.g. DEAE-Sephadex and DEAE- cellulose) have been used successfully as adsorbents for enzymes in industrial processes. By using the principles of affinity chromatography, more specific techniques for adsorption of enzymes have developed. Enzymes have been adsorbed through hydrophobic interaction onto polymers substituted with aliphatic side chains [9–11]. In other cases, allosteric enzymes have been immobilized through binding to immobilized effector molecules [12, 13].

2.2.2 Covalent Linkage

As a consequence of the covalent coupling of a protein to a solid support, one or more amino acid side chains of the protein are modified. This modification must not, of course, occur on amino acids which are essential for enzyme activity. Since many different amino acids can be utilized for coupling, it is in most cases possible to retain enzyme activity after immobilization by choosing a convenient coupling reaction. The active site can also be protected by carrying out the immobilization in the presence of a substrate or competitive inhibitor.
The derivitization of these functional groups on the protein should preferably be carried out in aqueous media under mild conditions. It is also desired, in order to mini-

mize side reactions, that the coupling reactions be relatively specific for one kind of functional group. Most of these reactions fall under one of the following major categories:

1. peptide bond formation
2. alkylation
3. diazo linkage
4. isourea linkage to BrCN-activated carbohydrates.

Before coupling can take place, however, functional groups on the carrier must in most cases be activated. Carboxyl groups can be activated by making azide or anhydride derivatives or by utilizing condensing agents such as carbodiimides or Woodward's reagent. These activated carboxyl groups react readily with primary amino groups on the protein to yield peptide bonds.

Alkylation of free amino groups as well as phenolic or sulfhydryl groups can be effected by using a polymer possessing a functional group such as a halogen.

Diazo coupling involves the preparation of an intermediate polydiazonium salt of the carrier and reacting it with the enzyme. Mainly tyrosyl residues, but also lysyl and arginyl residues, have been found to be involved in this coupling.

Inorganic supports, such as porous glass and silica beads, can be activated by coupling with α-aminopropyltriethoxy silane and the resulting compound can be converted to the reactive isothiocyanate or triazine derivatives. Diazotization of aryl amine derivatives of these materials can also be employed.

One of the most widely used methods for enzyme immobilization has been the BrCN-activation of carbohydrates. In this activation reaction an unstable reactive cyanate is formed which reacts with primary amines on the protein to form mainly an isourea derivative, but N-substituted carbonates and N-substituted imidocarbonates are also formed [14].

2.2.3 Entrapment and Encapsulation

Immobilization by entrapment or encapsulation differs from other methods of immobilization in the respect that the enzyme molecules are free in solution, but restricted in the room by a gel lattice or a membrane. The structure of the entrapping polymer should thus be tight enough to prevent the enzyme from leaking out and at the same time be loose enough to allow substrate and product to penetrate. Consequently, only reactions involving relatively small reactants may be carried out successfully by using preparations of entrapped enzyme.

The most widely used entrapment technique has been the occluding of enzymes or microbial cells within cross-linked polyacrylamide. In "block" polymerization, acrylamide is polymerized in the presence of a cross-linker in an aqueous medium containing the dissolved enzyme. The resulting gel block can be mechanically dispersed into particles of defined size. Bead polymerization is performed in a two-phase system. An aqueous solution containing the enzyme and the acrylic monomer is dispersed in a hydrophobic phase (i.e. toluene or chloroform) and subsequent polymerization of the emulsion yields well defined spherical beads [15, 16].

Entrapping enzymes in porous fibers can be achieved in the following way [17]. An organic solvent (not miscible with water) containing a fiber-forming polymer is emulsified with an aqueous solution of the enzyme and this emulsion is extruded through fine holes into a coagulant which precipitates the polymer in filamentous form. The most commonly used polymer is cellulose triacetate, but also other cellulose derivatives and poly-γ-methyl-L-glutamate can be used [17].

Finally, enzymes can be encapsulated within semipermeable nylon microcapsules [18] or within liposomes [19]. Microencapsulated enzyme preparations have yet only received attention for their potential use in medical applications.

2.2.4 Cross-Linking

Enzymes can be immobilized by intramolecular cross-linking into large aggregates by using bi- or multifunctional reagents. Even though this method is relatively simple, it has not been used to any great extent due to difficulties in controlling size and mechanical properties of the aggregates.

Cross-linking of the enzyme, however, has proven valuable in combination with other immobilization techniques. Enzymes adsorbed on solid supports have been cross-linked to increase the stability of the immobilized enzyme preparation and thus minimize leakage of the enzyme from the support [20, 21]. Cross-linking of entrapped enzymes has been carried out to improve stability [22].

In addition, intramolecular cross-linking between the enzyme of interest and an inert protein such as collagen [23], gelatin [24], or chitin [25] has received recent attention. This method is, in fact, a form of covalent coupling to an insoluble support.

In each of these cases the generally used cross-linking agent is glutaraldehyde, a dialdehyde that reacts with primary amines to form stable linkages. The exact nature of this linkage has not yet, however, been established with certainty. It has been suggested that the reaction probably involves addition of the amino group to ethylenic double bonds of α, β-unsaturated oligomers of glutaraldehyde [26].

2.3 Characteristics of Immobilized Enzymes

The properties of an immobilized enzyme system can be quite different from those of the corresponding soluble enzyme system. The changes can be attributed to alterations of the enzyme molecule itself or to the physical and chemical nature of the carrier. Furthermore, the transformation of the enzyme from a homogeneous to a heterogeneous catalyst can impose changes in catalytic properties.

2.3.1 Stability of Immobilized Enzymes

One of the most important features of an immobilized enzyme used in a large scale operation is the stability. Most enzymes when immobilized show a higher stability than their soluble counterparts. This increased stability could be the result of reduced conformational inactivation as well as reduced attack by reactive solutes due to steric shielding.

Different types of stability are distinguished for an immobilized enzyme. Of these, operational stability is the most relevant engineering parameter. Increased stability does not only decrease the cost of enzyme, but also reduces the cost of operation since fewer re-loadings of the reactor are needed. Operational stability is expressed as half-life, which is the time required for loss of 50% of the initial activity. The half-life is often obtained by extrapolation from rather limited data. Such a procedure can be disastrous unless very careful evaluation of the limited results is carried out. Operational stability can also be expressed, perhaps more accurately, with a rate equation incorporating a deactivation constant.

Storage stability becomes important when the enzyme preparation is used only intermittently or during prolonged storage in a stand-by capacity. If good storage stability is observed, large batches of immobilized enzyme can be produced and thus the cost of preparation will be reduced.

Immobilized enzymes can also show changes in thermal and pH stability as well as stability against denaturating agents such as urea.

2.3.2 Kinetic Behavior

Upon immobilization of an enzyme to a solid support also the kinetic pattern of the reaction is altered considerably. The basic kinetic parameters K_m (the Michaelis constant) and V_{max} (the maximum reaction velocity) are changed and the apparent values of these parameters are only observed. These changes are caused by steric, microenvironmental, and diffusional effects.

Steric limitations on the expression of enzyme activity are especially observed when the immobilized enzyme is catalyzing a reaction involving a high molecular weight substrate. It has been shown that different peptide maps are obtained when a particular protein substrate was hydrolyzed by either a soluble or an immobilized protease [27, 28]. In a similar way immobilized α-amylase gave a different product composition than the soluble enzyme when amylose or amylopectin was hydrolyzed [29]. In the case of α-amylase, a primarily endohydrolytic enzyme in free solution behaved like an exohydrolase after immobilization. Thus, the kinetic behavior can be modified to such an extent that a completely different product composition is obtained by the action of an immobilized enzyme.

The environment in the vicinity of an immobilized enzyme often differs from that in the bulk phase. If the enzyme is bound to a polyelectrolytic support, electrostatic interactions will lead to an unequal distribution of ions between the carrier phase and the bulk phase. Within a polyanionic carrier, the concentration of positively charged ions will be higher than in the external solution. Consequently, the pH inside such a polymer will be lower than in the surrounding media and the pH-activity profile of the immobilized enzyme will be shifted toward a higher value. In a polycationic polymer, the opposite situation prevails, i.e. the apparent pH-optimum for the immobilized enzyme is lower than for the native enzyme.

The apparent K_m can be similarly influenced if a charged substrate is processed by an enzyme bound to a polyelectrolyte. Identical charges on the substrate and the polymer could result in a higher apparent K_m due to repulsive electrostatic forces; if they are of opposite charge, a lower K_m could be observed. Electrostatic effects can often be reduced, even eliminated, by adjusting the ionic strength of the buffer.

Diffusion can be separated into external or bulk diffusion and internal or pore diffusion for a heterogeneous catalyst such as an immobilized enzyme. External diffusion resistance results from the occurrence of an unstirred layer, known as the Nernst diffusion layer, of fluid around the polymer particle. The thickness of this layer is inversely proportional to the velocity of the surrounding solution. Thus, by increasing the agitation rate in a batch process or the laminar flow rate in a packed bed process, the external diffusion resistance can be reduced, resulting in a lower apparent K_m.

Internal diffusion resistance substantially reduces the expressed activity of the immobilized enzyme. This effect is quantitated by an effectiveness factor, which is defined as the ratio of the actual reaction rate and the rate expected without diffusional resistance. This factor is inversely proportional to particle size. To eliminate internal diffusion resistance completely very small particles would be needed, and this would cause severe pressure-drop problems in industrial scale packed bed reactors. In many practical applications of immobilized enzymes, internal diffusion is consequently likely to be the rate controlling step. Internal diffusion resistance of the product can result in changes of the microenvironment, which may influence the kinetic behavior of the bound enzyme in a similar way as described above.

2.3.3 Multienzyme Systems

There are many enzymatic conversions of practical interest that involves the action of sequential enzymes. When two or more of such enzymes are immobilized on the same support, a very efficient system is obtained. This efficiency is particularly expressed in the initial phase of the reaction [30]. Due to diffusional limitations, the product from the first enzyme accumulates in the microenvironment of the bound enzymes and the second enzyme sees a relatively high concentration of substrate after a short reaction time. In contrast to the soluble system, the efficiency of the immobilized system is increased with an increasing number of sequential enzymes [31].

In practice, this microenvironmental effect can often be explored to produce efficient catalysts. First, co-immobilization of sequential enzymes will allow a higher flow rate in a packed bed reactor than if the enzymes were immobilized separately. Secondly, if, for instance, the first enzymatic reaction is thermodynamically unfavorable, the second enzyme will shift the equilibrium so a higher reaction rate is obtained [32]. Thirdly, if the enzymes are working intermittently on the same macromolecular substrate the proximity of the enzymes will allow a much faster reaction rate [33].

It should, however, be pointed out that co-immobilization is not to be recommended in all cases. When a feedback inhibition of enzymes occurs within the utilized reaction sequence, it is likely that more efficient catalysis will be obtained by separate immobilization of the enzymes. Likewise, when immobilized sequential enzymes show very different operational stabilities, immobilization in separate reactors can eliminate problems arising due to replacement.

2.3.4 Whole Cell Systems

When whole microbial cells are immobilized and used for various substrate transformations, still another diffusion barrier is introduced, i.e. the cell membrane. The diffusivity of this membrane restricts substrates to a small molecular weight. This drawback is compensated by several advantages such as the elimination of enzyme purification costs and in many cases increased enzyme stability. Furthermore, whole cell system can be used as single enzyme [34, 35], multienzyme [36], or whole pathway [37] systems and intracellular metabolites, e.g. coenzymes, can be utilized for synthetic purposes.

A restriction in the applicability of whole cell systems is the occurence of side reactions. When the difference of reaction rate between desirable and undesirable reactions is significant, a relatively good yield of product can be obtained by optimizing the reaction conditions [38]. In other cases, the problem can be solved by a selective inactivation of the unwanted side reaction [39].

2.4 Reactor Design and Performance

Reactor design for a process catalyzed by an immobilized enzyme does not differ fundamentally from the design of a conventional chemical process involving a heterogeneous catalyst. A number of different types of reactors have been used with immobilized enzymes in laboratory and industrial scale. Classification of these may be done according to the mode of operation and flow pattern.

2.4.1 Batch Reactors

Batch reactors are the most commonly used type of reactor when soluble enzymes are utilized as catalysts. In most cases no recovery of the enzyme is attempted. On the other hand, when expensive immobilized enzymes are used in a batch operation, an additional separation is required to recover the enzyme preparation. During this recovery procedure appreciable loss of immobilized enzyme can occur as well as inactivation of the enzyme due to repeated recovery cycles. It is, therefore, obvious that batch reactors have a rather limited potential in industrial immobilized enzyme catalysis. Futhermore, the use of a simple batch reactor does not take advantage of potential continuous operation, a major feature of immobilized enzyme systems.

2.4.2 Continuous-flow Reactors

Continuous reactors can be grouped into two basic types: the continuous-flow stirred-tank reactor (CSTR) and the plug-flow reactor (PFR). These reactors are fundamentally different and represent two extremes, i.e. complete mixing in the tank reactor and no mixing at all in the PFR. Consequently, conditions within the CSTR are the same as the outlet stream, while in the PFR the conditions vary with length from inlet to outlet. While the CSTR operates under uniform conditions of low substrate and high product concentrations, the conditions of the PFR are never uniform. In the latter case, the average conditions will result in high substrate and low product concentrations.

A nearly ideal CSTR is readily obtained in practice, but an ideal PFR is very difficult to construct. Several factors can give rise to deviations from a plug-flow pattern in a packed-bed reactor. Temperature and velocity gradients normal to the flow direction and substrate diffusion in the axial direction are the most frequently occuring complications, and even small deviations from the idealized flow pattern can alter the kinetics of the reaction considerably. In practice, tracing experiments should be carried out to estimate the approximation of plug-flow conditions prevailing in a packed-bed reactor before kinetic constants are determined.

In a fluidized-bed reactor, which will provide a degree of mixing intermediate to the two extremes described above, the substrate solution is passed upward at a velocity high enough to expand the bed of immobilized enzyme. Traditionally, this reactor configuration has been used when excellent heat and mass transfer characteristics are required. Although such requirements are not likely to occur in immobilized enzyme systems, fluidized-bed reactors do provide advantages of excellent flow rates and low pressure drop.

Variations of these basic types of reactors have been developed for specific purposes. In recycling reactors, a portion of the outflow is recycled and mixed with the inlet stream of the reactor. This procedure permits operation at relatively high flow rates, which minimize mass transfer resistance (external diffusion) in the reactor. The desired conversion is obtained by an appropriate recycling process. Finally, it appears to be possible to use slurry [40], trickle-bed [41], and slug-flow reactors [42] in special applications of immobilized enzyme systems.

2.4.3 Choise of Reactor Type

In designing a process, one of the first decisions which has to be made is whether the reactor will operate batchwise or continuously. It is quite obvious from what has been discussed earlier that in most cases a continuous process is preferred. However, the higher capital investment that is needed for a continuous process must be compared to the cheaper batch process with its higher labor cost. It should also be pointed out that a batch reactor has a broad flexibility for use in various processes, while a continuous process often requires a more specific reactor design.

Several factors will influence what type of continuous reactor will be chosen. Kinetic considerations play an important part in this choise. In general, the packed-bed reactors have intrinsic kinetic advantages over the stirred-tank reactors for most reaction types. In a CSTR, the average reaction rate is lower than in a packed-bed reactor due to the different operational concentration of substrate. In the case of a substrate inhibited reaction it is quite obvious that a tank reactor would be superior to any other kind of reactor.

The form and characteristics of the immobilized enzyme preperation must also be considered. In a stirred tank reactor the risk of loss of enzyme activity due to disintegration and/or solubilization of the support through mechanical shearing is much higher than in a packed-bed reactor. Thus, only relatively durable preparations of immobilized enzyme should be used in a CSTR. On the other hand, preparations of immobilized enzyme on very small particles can result in unacceptably high pressure drop and clogging problems when used in a packed-bed reactor. In this case the complications can be overcome by the utilization of a fluidized-bed reactor.

Operational requirements are still another factor which has to be taken into account. For instance a batch reactor is more suitable for reactions that require high oxygen transfer or the addition of base or acid for control of pH.

Finally, reactant characteristics can dictate the choice of reactor type. Particulate and colloidal substrates and products are processed preferably in fluidized-bed or tank reactors, where no plugging of the reactor is likely to occur, as would be the case in a packed-bed reactor.

In conclusion, there are no simple rules for choosing the reactor type for a specific process. As has been outlined in this section, a number of factors will influence the choice of reactor type and the advantages and disadvantages of a particular reactor must be analyzed in detail.

3 Applications in Food Industry

3.1 Sugar Industry

The sugar industry is today the largest consumer of industrial enzymes. In particular, enzymes are used for the production of glucose from starch and invert sugar from sucrose as well as for the isomerization of glucose to fructose. The interest in immobilized enzymes for these and other processes in the sugar industry is significant. As will be discussed below this interest has led to the development of some commercialized processes involving immobilized enzymes.

3.1.1 Hydrolysis of Polysaccharides

Glucose is mainly produced by the enzymatic hydrolysis of starch according to the reaction:

$$\text{starch} \xrightarrow{\alpha\text{-amylase}} \text{dextrins} \xrightarrow{\text{glucoamylase}} \text{glucose}. \tag{1}$$

The first step, liquification, is carried out with α-amylase in soluble form. This process has during recent years largely replaced acid hydrolysis. In the second step, saccharification, dextrins are hydrolyzed to glucose by glucoamylase. At present also this reaction is carried out with soluble enzyme.

Because α-amylase and glucoamylase are probably the most used enzymes in industry, significant research has been conducted on their immobilization, in particular glucoamylase. These enzymes are, however, relatively inexpensive, and therefore the immobilized enzyme must provide distinct economical or technical advantages.

α-Amylase from *Bacillus subtilis* has been immobilized on BrCN-activated carboxymethyl cellulose and used in a stirred tank reactor in the hydrolysis of wheat starch [43]. The productivity of the immobilized enzyme was lower than that of the soluble enzyme during the initial phase of the conversion, but because of thermal inactivation of the soluble enzyme, the relative activity of the immobilized enzyme was enhanced. An interesting observation is the distinct difference in action pattern, i.e. the immobilized α-amylase produced relatively more glucose and maltose than the soluble enzyme. A possible explanation for this effect relates to the limited diffusion of larger substrate molecules. Smaller fragments of the polysaccharide can readily diffuse into the interior of the polymer where they are completely hydrolyzed. It was shown that no external diffusion control existed and that the immobilized α-amylase could be used for several consecutive batches.

Glucoamylase has been immobilized covalently to various supports [44–47] and also entrapped in fibers [48]. Most studies, however, have not provided sufficient information for the evaluation of industrial applicability. One of the main advantages of immobilized glucoamylase would be to reduce the reaction time needed for hydrolysis. The 75 h duration required for the conventional batch process can be reduced to perhaps less than 1 h for the immobilized process. As a result fewer side reactions, e.g. polymerization of glucose to α-(1,6)-linked polysaccharide, should occur and the purification of the final product can be made easier and less expensive.

Glucoamylase immobilized on a copolymer of phenylene diamine and glutaraldehyde was operated in a plug-flow reactor (bed volume approximately 2 l) continuously at 48 °C for 3 months without any loss of activity [49]. When the reactor was fed with a α-amylase treated starch solution (DE 42) at a flow rate of SV = 0.4 a hydrolysis of 97% was achieved.

Immobilized glucoamylase has also been studied on a pilot plant scale [50]. The immobilization of the enzyme was actually carried out in the packed column by recycling a solution of glucoamylase through a bed of glutaraldehyde treated alkylamine silica beads. In this way 1.6 kg of glucoamylase was bound to 14.4 kg carrier, corresponding to an approximate 28 l bed volume. After immobilization, 45% of the glucoamylase activity was recovered. Investigations showed that the final DE-value and glucose concentration of the product not only were dependent on the residence time in the column, but also on the initial DE-value of the feed dextrin solution. The column was operated continuously for 70 days at 38 °C with 30 wt-% dextrin feed at flow rates of 250 and 500 kg per day, to produce a final product with a dextrose equivalent DE of 92-93.

On basis of the results from the pilot plant studies, an estimation of processing cost for a plant producing 45 million kg of glucose per year was made. The proposed reactor system contained a number of parallel columns which were maintained at constant conversion by decreasing the flow rates as the activity of the immobilized enzyme decreased. When the use of a column was estimated as two half-lives, it was concluded that at least seven columns would be needed to keep the production rate variation within 10%. Furthermore, it was concluded that operation at relatively low temperatures with two or three half-lives utilization of enzyme would be necessary to compete with the estimated cost of the soluble enzyme process if installed in a new plant.

When the carrier is recycled, its operational life will influence the production cost. Relations between this cost and carrier and enzyme stability have been demonstrated with data on glucoamylase immobilized on porous glass [51]. In this case, a plant designed to produce 4.5 million kg of glucose per year was chosen. The production cost was also in this study found to be sensitive to operation temperature and the optimum temperature was low, close to room temperature. At such a low temperature, however, the cycle times are very long and therefore a practical design was made by limiting cycle time to an upper limit and calculating the temperature for which this time was optimal. Futhermore, it was concluded that the total cost is not very sensitive to large deviations from optimum values of cycle time and number of cycles.

In practice such low temperatures as have been considered in these studies are not appealing. Manufacturers of glucose seem not to be interested in processes operated below 55–60 °C because of the risk of growth of microbial contaminants. Thus, a more heat

stable glucoamylase is needed before the commercialization of immobilized glucoamylase is a reality. Furthermore, it is not likely that such a process will be considered
as long as present production capacity is sufficient to supply the required amount of
glucose.

β-Maltose can also be produced from starch by the action of β-amylase, which hydro-
lyses α-1,4-glycosidic linkages from the non-reducing ends of starch chains. For complete
hydrolysis of starch to maltose, an additional enzyme such as α-amylase or pullulanase,
is required. β-Amylase is used in brewing, distilling, and baking industries to convert
starch to fermentable sugars. High maltose syrups are used in the confectionary industry because they are non-hydroscopic and do not crystallize as readily as high glucose
syrups.

Production of maltose syrups by an immobilized two-enzyme system has been proposed [33,52,53]. A relatively efficient system was obtained by co-immobilizing β-amylase and pullulanase, a debranching enzyme, in the same support, a cross-linked copolymer of acrylamide and acrylic acid. Since these two enzymes work intermittently on
the same substrate, a pronounced advantageous microenvironmental effect can be expected for the immobilized system (see Section 2.3.3). Increased operational stability
was observed compared to the soluble enzymes and operation in a packed bed reactor
was favored.

In a similar study, β-amylase was immobilized together with an α-1,6-glucosidase through
adsorption to active coal [54]. Both enzymes from *Bacillus cereus* were quantitatively
adsorbed to the support. This immobilized two-enzyme system hydrolyzed starch at
50 °C to a product composition of 90.5% maltose, 7.5% maltotriose, and 2% other oligosaccharides in a batch process.

The enzymatic hydrolysis of cellulose to glucose is probably one of the most exciting
new developments in enzyme engineering. Cellulose is the most abundant organic material and unlike other potential energy sources it is renewable. Very large quantities of
cellulose are wasted and the development of simple, efficient, and economical processes
for the conversion of such waste cellulose to glucose is of global interest. Glucose produced in such a way, if not used as sweetener, may serve as an energy source for single
cell protein production or it can be converted to other valuable chemicals (e.g. ethanol).
The hydrolysis of cellulose is catalyzed by cellulase (from e.g. *Trichoderma viride*),
which is a complex mixture of several enzymes with various hydrolytic activities. Development of a process involving addition of soluble cellulase to a suspension of cellulose
has reached a prepilot plant stage [55]. Some attempts have been made to utilize immobilized cellulase for the production of glucose from cellulose [56,57]. Since cellulase
consists of a mixture of enzymes with different physical properties, a fairly general
method of immobilization, which does not seriously inactivate any of the components
of cellulase, has to be used. Thus, cellulase was coupled to collagen beads by cross-
linking with glutaraldehyde [56]. In order to avoid plugging of the column by the
colloidal cellulosic substrate the enzyme preparation was used in a fluidized bed reactor. The substrate solution was recycled at an optimum flow rate through the column
and a quantitative hydrolysis was achieved. Upon immobilization a remarkable increase
in the stability of the enzyme was observed, i.e. from a half-life of 30 h for the native
to 21 days for the immobilized cellulase at 30 °C.

3.1.2 Isomerization of Glucose

High glucose syrup, which is produced by the action of α-amylase and glucoamylase on corn starch, is 70–75% as sweet as sucrose. By partial isomerization of glucose to fructose, the sweetness of this comparably inexpensive product can be increased to the same sweetness as the more expensive sucrose. Since there is no convenient chemical process available for this isomerization, syrups containing fructose are produced by recently developed enzymatic processes. In fact, all high fructose syrups (HFS) are produced by using immobilized glucose isomerase. This application is currently the largest volume use of an immobilized enzyme in the world. The total capacity of plants in USA and Europe can at present be estimated to be approximately 2–3 million tons of HFS per year. However, plants located within the European Economic Community have during 1977 experienced some difficulties due to political and agricultural reasons. In order to secure the production of domestic sucrose and accompanying labor, a high import tariff has been introduced on the inexpensive raw material, corn. It is therefore likely that the development of this industry in Europe will stagnate.

Several companies now offer a complete process or technology for production of HFS. Clinton Corn Products, one of the pioneers in the field, introduced an industrial process based on immobilized glucose isomerase (isolated from a *Streptomyces* species) in 1972 [58]. The enzyme was adsorbed on DEAE-cellulose and used in a continuous computer controlled process. The reactor consisted of several shallow beds placed in series with each bed containing material at a different stage of enzyme lifetime. The isomerization was carried out at 60 °C, pH 7.0–7.5. The substrate was a 73% glucose solution of 30–50% solids and the product contained 42% fructose.

The process developed by Novo Industries [59–62] for the production of HFS appears to be the most widely employed with seven plants in the USA and probably as many in Europe, Japan, and South Korea, with design capacities of up to 1200 tons per day (as dry substance). The total capacity of these plants can be estimated to be between 1.5 and 2 million tons per year. The glucose isomerase is obtained from *Bacillus coagulans* and it is immobilized simply by cross-linking the crude cell homogenate with glutaraldehyde. The cross-linked cell mass is subsequently made into pellets by extrusion. The first commercial process, which was introduced in 1974, was operated batch-wise. Since then, the mechanical properties of the immobilized glucose isomerase preperation have been improved and, today, most plants are utilizing packed bed reactors. Actually a packed bed reactor has some very distinct advantages over a batch reactor. In Tab. 2 some process parameters of the different reactor designs are compared. In the batch process the reaction conditions are not optimal because under these conditions, color and by-products are formed due to the long reaction time required. In the packed bed process, on the other hand, the reaction time is relatively short and optimal conditions for the isomerization can be utilized. In summarizing, the main advantages of the packed bed reactor design in this system are as follows: low cost of additives and purification; no requirement for cobalt; smaller reactor volume; significantly lower enzyme consumption, because it is possible to keep both pH and temperature in the optimum range without formation of by-products.

For the production of 100 tons per day (as dry substance) of HFS containing 42%

Table 2. Comparison of batch and continuous processes for isomerization of glucose to fructose (in the Novo system [61, 62])

Parameter	Batch process	Continuous Process
Relative reactor volume	25	1
Relative enzyme consumption	2	1
Reaction time (h)	20	4
pH	6.6–7.0	8.0–8.5
Temperature ($^{\circ}$C)	60–65	60–65
Concentration of activators: Mg^{2+} (M)	$8 \cdot 10^{-3}$	$4 \cdot 10^{-4}$
$\qquad\qquad\qquad\qquad\qquad$ Co^{2+} (M)	$3.5 \cdot 10^{-4}$	0

fructose, the following practical design has been proposed [61,62]. Six reactors, each with a bed volume of 2.2 m^3, are arranged in two parallel lines with three reactors in each line, as outlined in Fig. 2. Since the recommended lifetime is 1050 h (25% of the initial enzyme activity is remaining) and each reactor contains material at a different stage of enzyme lifetime, the enzyme has to be changed in one reactor every 175 h. The fresh enzyme reactor is always introduced as the last in its line in order to be heated to operating temperature without affecting the operation of the other reactors. With this suggested arrangement, the flow rate would not have to be varied more than 10% from average in order to obtain a constant degree of conversion. If, on the other hand, all the enzyme had been packed into one large reactor, the flow rate would have to be varied from about 170% to about 40% of the average during the enzyme lifetime of approximately six weeks. The flow rate is adjusted seperately for each line of reactors a few times per day to obtain 42% fructose at the outlet.

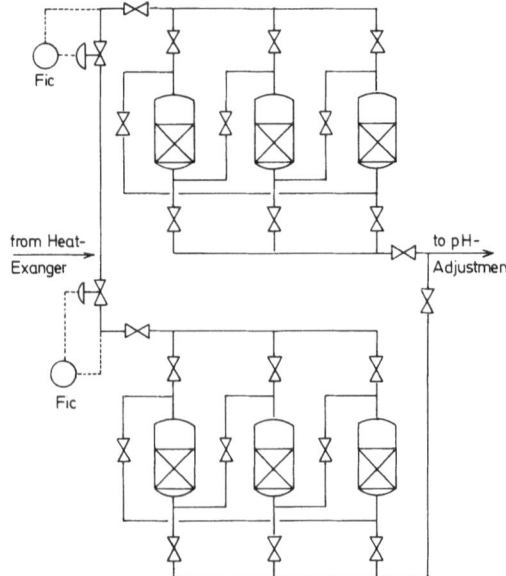

Fig. 2. Arrangement of reactors for production of 100 t per day of high fructose syrup (from [61])

Gist-Brocades (Holland) manufactures an immobilized glucose isomerase derived from a strain of *Actinoplanes missouriensis* [63]. The mycelium-fixed enzyme is entrapped in gelatin and cross-linked with glutaraldehyde. This system of immobilization was chosen because the carrier material and coupling reagents are permitted in food stuffs. The entire procedure of immobilization is continuous and applicable to other enzymes applied in food processing. The enzyme preperation consists of uniform spherical particles of one millimeter diameter. The manufacturer recommends the use of this glucose isomerase preparation for production of HFS in either a discontinuous batch process or a continuous process in a column with upward (expanded bed) or down- ward flow (fixed bed). The half-life of the enzyme preparation is reported to be 430 h for the batch process with cobalt present and 500 h for the column operation in the absence of cobalt. In both processes a 42% fructose syrup is obtained by operating at 65 °C, pH 7.0–7.5.

Other processes for HFS production are available from many companies. Corning Glass Works has developed a process based on glucose isomerase from a strain of *Strepto- myces* [64]. The enzyme was adsorbed on controlled-pore alumina, which had advan- tages over other inorganic carriers. R.J. Reynolds Tobacco Co. utilizes a strain of *Arthrobacter* for production of glucose isomerase [65]. The cells are aggregated and further processed to yield a cell material which can be used in a fixed bed reactor. Finally, Baxter Laboratories has a process based on enzyme from *Streptomyces phaeochromogenes*, which is immobilized on an anion exchange cellulose support [66]. The list of immobilized glucose isomerase preparations can be extended considerably. For instance, purified glucose isomerase has been immobilized on DEAE-cellulose [67], porous glass [68, 69], phenylformaldehyde resins [70], and also entrapped in cellulose fibers [71] and hollow fibers [72]. Whole cells with isomerase activity have been immo- bilized by entrapment in cellulose acetate [73] and cellulose fibers [74], as well as poly- acrylamide [75] and collagen membranes [76]. The potential use as industrial catalysts of most of these preparations appears, however, to be rather limited, especially since there are a number of processes, as described above, already on the market.

An interesting scaling-up calculation has been conducted on the basis of a reactor design equation obtained from laboratory scale reactors [76]. The catalyst, consisting of whole cells of a *Bacillus* species incorporated in collagen membranes, was used in the form of chips in a packed bed reactor. The process was designed for a capacity of 135 million kg of HFS per year and it was concluded that 21 reactors operating in a sequential mode were required. Each reactor contained 860 kg of catalyst with a life- time of 5400 h (3 half-lives). Furthermore, it was assumed that the recharging of the reactors was staggered in such a way that the average activity of the entire catalyst mass at any time was very close to constant. The feed solution consisted of 50% solids (73% glucose) and the final product 71% solids (50% glucose, 42% fructose). It was conclu- ded that this isomerization process is attractive for commercialization since the net return on investment was calculated to be 16.2%.

3.1.3 Hydrolysis of Sucrose

Sucrose can be hydrolyzed to invert sugar by acids or enzyme. The product, consisting of a 1 : 1 mixture of glucose and fructose, is somewhat sweeter than sucrose. It is used in food and confectionary products as a humectant to hold moisture and prevent drying out and in the brewing industry. The enzymatic hydrolysis, which is catalyzed by invertase (β-fructofuranase), yields a product of higher purity than acid hydrolysis. Even though the production of invert sugar with soluble enzyme is relatively inexpensive, attempts have been made to immobilize the enzyme for continuous production of invert syrups. Invertase is a relatively stable enzyme and immobilization results in preparations with long lifetimes.

Invertase from yeast was entrapped in cellulose triacetate fibers and evaluated for industrial use [77]. In immobilized form the enzyme was found to have satisfactory activity and stability under operational conditions, especially in high sucrose concentrations where the free enzyme showed a severe substrate inhibition. It was concluded that the observed differences between free and entrapped invertase were due to diffusional effects.

In a study on invertase from *Candida utilis* immobilized on porous cellulose beads, it was similarly concluded that diffusion hindrance actually helps to minimize substrate inhibition [78]. Under the condition used, both external and internal diffusion barriers contributed to this effect. Thus, diffusion barriers, which normally are disadvantageous for the practical utilization of immobilized enzymes, in these cases, influence the kinetic behavior of the immobilized invertase in a positive manner.

Invertase, entrapped in polyacrylamide, retained 90% of the initial activity after continuous operation for 450 days at 30 °C [79]. High sucrose concentrations (30—50%) resulted in increased flow resistance in a column operation due to increased viscosity of the substrate solution. Therefore, a batch process was recommended for the hydrolysis of solutions of high sucrose content.

These and other studies on immobilized invertase indicate the interest and potential use of such preparations for the continuous production of invert sugar on a industrial scale. However, so far no scaling-up of a process has been reported, and before commercialization more extensive studies on reactor design and reaction parameters are required.

3.1.4 Hydrolysis of Raffinose

Raffinose, an α-galactoside, interferes with the crystallization of sucrose from sugar beet molasses. The yield of sucrose can be increased by the hydrolysis of raffinose by α-galactosidase. This increased yield is partly due to the actual formation of sucrose from the hydrolysis of raffinose and partly due to the elimination of the retarded crystallization of sucrose. For the development of an enzymatic process to be used in the sugar beet industry, a microorganism producing significant amounts of α-galactosidase and very little or no invertase was required. The mold *Mortierella vinacea var.* raffinoseutilizer fulfilled these requirements and was therefore selected as a good source of the enzyme [80]. When the mold was grown under certain controlled conditions mycelial pellets of uniform size (20—30 mesh) containing α-galactosidase was produced. These pellets are conveniently used as a preparation of immobilized α-galactosidase in a continuous process.

An industrial process involving pellet-bound α-galactosidase was introduced in 1968 by Hokkaido Sugar Co., Ltd., Japan at the Kitami factory [81]. A similar process was designed for Great Western Sugar Co., USA, at the Billings factory. The process was ready for use in late 1973 but a delayed FDA approval of the process postponed the start up until late 1974 [81].

The enzyme reactor consists of a number of U-shaped open vessels each equipped with agitators and a replaceable screen at the top for removal of catalyst from the product stream. The molasses to be treated in the reactor is continuously adjusted to 29–31 °Brix and a temperature of 48–52 °C with warm water and pH 5.0–5.2 with sulfuric acid. The optimum pH for the α-galactosidase reaction is 4.0, but is maintained at 5.2 to avoid inversion of sucrose. The reaction time is 1.5–2.5 h but up to 4 h can be used without trouble if molasses of higher raffinose content is treated. The enzyme pellets are used for about 25 days before they are discharged. Fresh enzyme preparation is added in a chamber near the inlet of substrate and, upon aging, is transferred towards the product outlet. About 30 kg of enzyme pellets are used to hydrolyze 1 ton of raffinose when the enzyme reaction is continued for 100 days, and 60% of the trisaccharide in the molasses is converted.

Since the introduction of the enzyme process, the capacity of beet processing increased by 11–12% and the sucrose extraction increased from 87.8% to 90.7%. At the same time the amount of molasses that had to be discarded was reduced from 1.07%, based on beet quantity, to 0%. The economical value of these improvements is quite obvious and may explain the commercial success of this process.

The enzyme has also been insolubilized within the cells of *M. vinacea var.* raffinoseutilizer with glutaraldehyde [82]. In a batch reactor, beet sugar solutions containing raffinose were treated with the cells at 50 °C. After more than 250 cycles over a 3 week period the treated cells retained 72% of original activity whereas the untreated cells retained only approximately 5%.

Pellets were made from cells of *Absidia lignierii* containing α-galactosidase by using a pellet machine [83]. Diluted beet juice containing 3.54% raffinose was passed through a column packed with the pellets at 60 °C and at a flow rate of SV = 0.5. During 30 days of operation, the α-galactosidase activity decreased to 80% of the initial activity. Finally, α-galactosidase from *Bacillus stearothermophilus* was immobilized on a nylon carrier and used in a packed bed reactor for hydrolysis of raffinose in diluted beet molasses [84]. The reactor was designed to hydrolyze 95% of a 1.5% raffinose solution with a retention time of 1 h. To avoid clogging, the diluted molasses was filtered before it was passed through the column.

Immobilized α-galactosidase has also been considered a potential catalyst for the hydrolysis of α-galactosides present in soybean milk in a large scale operation (see Section 3.2.4)

3.1.5 Production of Gluconic Acid

Immobilized glucose oxidase has several potential applications in the food industry. It can be used in the production of gluconic acid from glucose. If a high fructose syrup or invert sugar is treated with glucose oxidase, a mixture of fructose and gluconic acid will be obtained, which is easily resolved by ion-exchange chromatography. Thus, a simultaneous production of pure fructose and gluconic acid is feasible. Glucose oxidase can also be utilized for the removal of glucose from foodstuffs, such as egg white (see Section 3.2.4) and also for the removal of oxygen in a preservative manner.

The products from the reaction catalyzed by glucose oxidase are a labile δ-glucono-lactone, which spontaneously hydrolyzes to gluconic acid, and hydrogen peroxide. The latter product is harmful to the enzyme and therefore it should be eliminated as efficiently as possible. This is conveniently achieved by the action of catalase, which converts hydrogen peroxide to oxygen and water. Glucose oxidase and catalase have been co-immobilized and studied in many laboratories [85—88]. It has been concluded that the stability of glucose oxidase is greatly improved when catalase is also present on the support. The actual activity of glucose oxidase is also influenced since the oxygen, which is consumed in the oxidation reaction, is reformed by the action of catalase. This recycling of oxygen can be expected to be relatively efficient due to microenvironmental effects.

Aspergillus niger is a very convenient source of these enzymes because they can be simultaneously purified and a desired ratio of enzyme activities is also obtained. This two-enzyme system has been immobilized on a copolymer of phenylene diamine and glutaraldehyde and studied in batch and continuous column reactors [49]. It is interesting to note that for the immobilized system, a lower temperature optimum than required for the soluble system was observed. This might be explained by considering a non-saturated concentration of oxygen in the microenvironment of the enzymes at higher temperatures. Aeration with oxygen gave a higher reaction rate than when air was used. Furthermore, a decreased K_m for glucose was observed for the immobilized system which probably was the result of a better supply of oxygen in the proximity of catalase. Thus, it appears that the rate limiting step is not diffusion of glucose, but rather the supply of oxygen. It was concluded that glucose oxidase, when co-immobilized with catalase, was an efficient catalyst for the production of gluconic acid and that it might find application on an industrial scale.

3.2 Dairy Industry

Dairy industry utilizes enzymes in various processes and new applications are under development. Many of these old and new processes are suitable for the application of immobilized enzymes. One of the most important new developments appears to be the use of immobilized β-galactosidase for hydrolysis of lactose in milk products. This and other processes involving immobilized enzymes will be discussed in the next sections.

3.2.1 Hydrolysis of Lactose

The hydrolysis of lactose to form glucose and galactose is of interest from several points of view. A significant percentage of the world population is intolerant to lactose and therefore may not drink milk without suffering diarrhea and gastrointestinal distress. The production of cheese gives rise to whey as a by-product, which contains relatively high concentrations of lactose. The utilization of this by-product is to some extent limited because the disaccharide tends to crystallize. Very large volumes of whey are therefore discharged every year causing waste disposal problems and the loss of valu-

able nutrients. Most of these problems can be eliminated or reduced by enzymatic
hydrolysis of lactose with immobilized β-galactosidase (lactase).

The large number of publications on the immobilization of β-galactosidase during
recent years reflect this great interest. The enzyme can be obtained from yeast such
as *Saccharomyces lactis,* molds like *Aspergillus niger,* or such bacteria as *Escherichia
coli.* β-Galactosidase from these and other sources have been immobilized on a variety
of supports including inorganic carriers [89–91], collagen [92], and agarose [93, 94] as
well as entrapped in polyacrylamide gels [95, 96] and cellulose triacetate fibers [97].
Whole microbial cells exhibiting β-galactosidase activity have been immobilized in polyacrylamide gels [98]. The choice of enzyme source is influenced by the application of
the preparation. Hydrolysis of lactose in acid whey, for instance, is conveniently achieved
with the enzyme from *A. niger* which has a pH-optimum of about 4.0, while for other
substrates of neutral pH, such as milk, the enzyme from *S. lactis* or *E.coli* is more
suitable.

Some of the preparations of immobilized β-galactosidase listed above have been evaluated for a large scale operation or investigated in a pilot plant. Probably, the first
pilot plant-sized lactose hydrolyzing process was reported by Snam Progetti in coope-
ration with the Dairy plant of Milan, Italy [99, 100]. Yeast β-galactosidase was entrapped in cellulose triacetate fibers and used to hydrolyze lactose in sterilized skim
milk. The process was operated batchwise by recycling the milk through a column
containing the fiber-entrapped enzyme for 21 h at 7 °C. After 50 cycles, 10000 l milk
had been processed and some conclusions were drawn. No problem with bacterial contamination was experienced if appropriate disinfection was carried out when required.
The loss of enzymatic activity was very low and after 50 cycles only 9% of the initial
activity had been lost. Furthermore, the use of fibers of higher activity led to a marked
improvement in hydrolysis thereby reducing the fiber to milk ratio. In fact, the results
of the pilot plant experiments were so encouraging that an industrial plant with the
capacity of 10 t of treated milk per day was built in Milan and was ready for use in
1977.

The hydrolysis of lactose in acid whey is of interest for the production of sweeter,
more soluble sugars that can be used in dairy products. With increasing cost of sweeteners, this possibility becomes more attractive and an estimation of the production
cost of sweeteners from whey has been carried out [91]. A plant design including an
ion-exchange step for prior demineralization of the whey and a concentration step
after the hydrolysis was considered. The size of the system was based on the use of
immobilized β-galactosidase with an apparent activity of 300 U/g at 35 °C. Preparations of similar or higher activity could be obtained by immobilizing the enzyme
from *A. niger* on porous silica. In the suggested operation, the enzyme column was
initially run at 35 °C and, as the activity of the immobilized enzyme decreased, the
temperature was raised to maintain the initial conversion level until 50 °C was reached.
Under these conditions the life-time of the enzyme preparation was calculated to be
559 days.

The process cost included labor and supplies, capital or equipment costs taken as 20%
annually, and the cost of immobilized enzyme. Total cost appeared to be 2–8 cents
per kg of hydrolyzed lactose, depending on plant size, cost of immobilized enzyme

and conversion level. With the two additional steps, i.e. demineralization and concentration, the overall cost was estimated to be 18—22 cents per kg, which was concluded to be competitive with the price of other sweeteners.

The process, which was developed by Corning Glass Works, has been tested on a pilot plant scale [101]. A column with the capacity of 6800 l per day of liquid whey was utilized. Before the whey was passed through the enzyme column dissolved protein was removed by ultrafiltration.

The hydrolysis of lactose in whey has also been carried out with β-galactosidase immobilized on alumina particles on a pilot plant scale [102]. The process was operated as a fluidized bed in a column of 3 inch diameter. The design of the pilot plant was very flexible with provision for prior ion-exchange and ultrafiltration steps and a concentration step subsequent to the hydrolysis. During some operational conditions, a layer of protein was formed on the enzyme particles, which, however, could be removed simply by sonication. Tracer experiments revealed that far from plug-flow conditions were prevailing in the reactor and therefore improvements of the process are affordable. Nevertheless, based on the pilot plant experiments, an industrial plant was designed. With the capacity set to 45000 kg of whey per day and the conversion to 70%, a column of 50 cm diameter and 7 m height would be required. The economics of the process depends of course on the number of different steps included. From economic analysis, it was concluded that the market value of the protein obtained in the ultrafiltration step was of great importance for the expected return on investments. In conclusion, an immobilized enzyme process for the hydrolysis of lactose in whey seems to be attractive and close to realization. In addition it should once again be pointed out that such a process would eliminate disposal problems connected with whey.

3.2.2 Cheese Manufacturing

Coagulation of milk can be divided into two distinct phases. In the primary or enzymatic phase, a proteolytic enzyme such as rennet or pepsin cleaves a phenylalanine-methionine bond of κ-casein leading to a metastable state of the casein micelle. In the secondary or nonenzymatic phase, which requires calcium, the milk gels. The temperature coefficients (Q_{10}) for these reactions are 2 and 10-12, respectively [103]. Consequently, coagulation of milk can be selectively retarded in an immobilized enzyme reactor by lowering the temperature to inhibit the nonenzymatic phase but allowing completion of the enzymatic phase. Upon warming the treated milk will curd. A continuous process for cheese manufacturing is thus feasible and since such a process offers advantages over the traditional labor-intense batch process, it may find application in commercial cheesemaking. In spite of many studies on the use of immobilized protease for milk-clotting [103—106], it appears that many problems still have to be solved before such a process is realized. Use of expensive carriers and complicated immobilization techniques, as well as low activities of immobilized enzymes and poor stabilities, are some of the problems experienced in these and other studies.

A promising system involving a cheap carrier (alumina), a simple immobilization technique (adsorption), and a relatively active enzyme (pepsin) has been developed for use in a fluidized bed reactor [107].

A look at some additional advantages of a continuous cheesemaking process based on an immobilized protease may explain the continuing efforts made to develop such a system. Shortage of milk-clotting enzymes may be eased by taking advantage of immobilized enzymes which can be re-used. Proteases, which are not suitable for use in soluble form because of excessive uncontrolled proteolysis, can be used in an immobilized state. The enzyme used in the current process was chosen for its milk-clotting capacity as well as for its influence on the cheese ripening process. An immobilized clotting enzyme would allow the use of a second protease selected for its beneficial effects on cheese ripening, thereby avoiding a compromise between milk-clotting and ripening capacities. Finally, a very flexible reactor design is possible when an immobilized protease is used.

3.2.3 Sterilization of Dairy Products

In the dairy industry, up to 0.05% H_2O_2 has been used to "cold pasteurize" milk being prepared for the manufacturing of cheese. After the pasteurization, the H_2O_2 is destroyed by catalase. The high cost of the enzyme has, however, limited the use of this technique. An immobilized catalase might prove useful to remove H_2O_2 from treated milk.

Investigators, studying immobilized catalase, have been discouraged by the fact that H_2O_2 inactivates catalase rapidly [108—110]. However, a report on the possible reactivation of the enzyme, in this case immobilized on collagen membranes, has to some extent revived interest in the use of immobilized catalase in the food industry [111].

An antibacterial agent is formed in the lactoperoxidase catalyzed oxidation of thiocyanate by H_2O_2. Milk contains high concentrations of peroxidase and moderate amounts of thiocyanate. The addition of H_2O_2 to milk leads to the activation of the lactoperoxidase system resulting in "self-pasteurization". Generation of H_2O_2 from lactose with an immobilized two-enzyme system has been suggested [112]. β-Galactosidase and glucose oxidase were co-immobilized on porous beads and used for the treatment of whey in a packed bed reactor. Good antibacterial effects were observed when the whey was passed through the column at a flow rate resulting in almost complete consumption of dissolved oxygen. An obvious advantage of the system is that no additional destruction of excess H_2O_2 is necessary.

3.2.4 Miscellaneous Applications

It has been known for a long time that treatment of milk with trypsin inhibits the development of an oxidized flavor on storage. Trypsin immobilized on porous glass has been prepared and utilized for the treatment of milk [113, 114]. A packed bed reactor was chosen to be more efficient than a fluidized bed or batch reactor. This reactor design, however, was not ideal since an increased pressure drop was observed due to the deposition of a fat layer at the top of the enzyme bed during operation. Raw milk was pumped through the reactor for 5 h daily for a total of 14 days. At an appropriate flow rate, all samples treated were effectively protected from oxidized flavor. No bacterial contamination problems were experienced during the entire course of the experiments. The immobilized trypsin showed good storage stability in 20%

ethanol and was not affected adversely by repeated contact with milk. It was concluded that the excellent stability would allow repeated and prolonged usage of the immobilized enzyme and thereby reduce the cost of the process.

A serious problem in the production of ultra-high temperature sterilized milk, which can be marketed at ambient temperature, is the occurence of a cooked flavor. This off-flavor is connected with the exposure of sulfhydryl groups due to the thermal denaturation of proteins in the milk. Re-oxidation of these groups would thus eliminate this problem. A sulfhydryl oxidase with broad substrate specificity has been isolated from milk and immobilized on porous glass [115]. The enzyme was immobilized on a carboxylated glass rather than the commonly used alkylamino form. There is evidence that milk proteins, which are anionic at neutral pH, adsorb to the alkylamino-derivative of the polymer leading to complications during the process. By utilizing an anionic polymer, adsorption can be reduced or eliminated. This approach may have general applicability to systems involving the treatment of milk. Preliminary studies have shown that the amount of reactive sulfhydryl groups in heated skim milk can be reduced considerably by passing the milk through a reactor containing the immobilized enzyme leading to the elimination of the cooked flavor. The reactor was operated daily for 3-5 h at 35 °C and after 10 days 70% of the initial activity remained, which was considered encouraging for further studies. Furthermore, a process based on the immobilized oxidase is the only realistic system since a soluble system requires a large amount of enzyme, which is difficult to obtain [116].

As mentioned earlier (see Section 3.1.5) an immobilized two-enzyme system consisting of glucose oxidase and catalase may be utilized for the removal of glucose from egg whites before drying. If the glucose is not removed, the dehydrated egg whites, which are used in cookie and cake mixes, become rancid and brown-coloured. At present the removal of glucose is accomplished by the addition of glucose oxidase to the milled egg whites. However, this enzyme, which is of intracellular origin, is relatively expensive and therefore an immobilized enzyme system would be beneficial. In a series of experiments using immobilized glucose oxidase and catalase in a continuous packed bed reactor, the operation was hindered by plugging and gel formation [117]. Precipitation and removal of mucin protein eliminated these complications, but this solution was considered unsatisfactory by the egg industry. Consequently, it appears that a batch operation would be more favorable.

The use of immobilized α-galactosidase for the hydrolysis of raffinose in sugar beet molasses was described above (see Section 3.1.4). There is also interest in this enzyme for the hydrolysis of the α-galactosides, raffinose, stachyose, and verbascose, which are considered to be partially responsible for the flatulence associated with dry bean products such as soybean milk [118, 119].

Mycelia of *Mortierella vinacea* were immobilized by entrapment in polyacrylamide gel, and were used in a fluidized bed reactor, operated at 50 °C, to hydrolyze oligosaccharides of soybean milk [118]. At less than practical flow rates, up to 60% hydrolysis of the saccharides was obtained. An alternative immobilization method, which could give a derivative of higher activity, would be necessary for satisfactory results.

Crude preparations of α-galactosidase, readily prepared from *Aspergillus awamori*, were used in a hollow-fiber reactor for the treatment of soybean milk or whey [119]. By recycling the substrate solution, complete hydrolysis of the oligosaccharides was obtained.

3.3 Production of Amino Acids

During recent years amino acids have become increasingly important as food supplements and medicinal agents. Amino acids are also used as cosmetic additives and as raw materials for the synthesis of other compounds. In most cases only the biologically acitve L-isomers of the amino acids are of interest in food and pharmaceutical industries.

Chemical and biosynthetic methods are the most useful for the production of L-amino acids. Chemical synthesis results in a racemic mixture of the L- and D-isomers, which subsequently needs to be resolved. By using biosynthetic methods, on the other hand, the L-isomer is specifically produced. L-Amino acids can also be produced by hydrolysis of proteins with proteases and peptidases.

Application of immobilized enzymes or microbial cells for the production of L-amino acids has proven to be very useful. In fact, this technology has been utilized in some cases for the commercial production of amino acids. Many other process involving immobilization of biocatalysts for production of amino acids appear to be close to commercialization.

3.3.1 Resolution of Racemic Mixtures

For the optical resolution of racemic amino acids, chemical, enzymatic, or biological methods are available. One of the most advantageous procedures is an enzymatic method which involves the enzyme aminoacylase. By the action of this enzyme, an acyl-D,L-amino acid is asymmetrically hydrolyzed and the L-amino acid is formed. The two products, which are formed in the enzyme reaction, can easily be separated by differences in solubility. The isolated acyl-D-isomer can be racemized and recycled to increase the yield of the L-isomer. The enzyme has a broad substrate specificity and is therefore very useful.

Tanabe Seiyaku Co., Japan, has used aminoacylase from *Aspergillus oryzae* for the industrial production of amino acids. A process involving immobilized aminoacylase has been developed in order to overcome some disadvantages experienced with a batch process involving soluble enyme [120]. Before an applicable system was obtained an extensive study on the immobilization of aminoacylase and its characterization was carried out. The development of the process is illustrative and therefore it will be reviewed in some detail.

A large number of different immobilization techniques were investigated and three systems with promising results were selected for further evaluation. These three preparations of immobilized aminoacylase, which were prepared by ionic binding to DEAE-Sephadex, covalent linkage to iodoacetyl cellulose and entrapment in polyacrylamide, were compared regarding enzymic properties. No significant differences between any of these preparations and native enzyme were observed except that DEAE-Sephadex-aminoacylase was somewhat more heat stable than the other preparations. The final choice was based on the following criteria:

(a) activity and stability of the immobilized enzyme
(b) ease and cost of immobilization
(c) possibility of regeneration of the column when the enzyme activity is lost.
By all these criteria immobilization of the enzyme on DEAE-Sephadex appeared to
be the most suitable for industrial use. Since no chemical reactions are needed for
immobilization, large quantities of carrier-enzyme complex can be prepared easily.
Furthermore, DEAE-Sephadex is the only support of those considered which can be
regenerated. In the preparation of the immobilized aminoacylase, 323 units of native
enzyme were used per ml of DEAE-Sephadex A-25 and the resulting complex had an
activity of 157 units per ml, corresponding to a coupling yield of 47%.
For the design of the enzyme reactor, the following factors were regarded as being
important:
(a) effect of flow rate of substrate on the reaction rate
(b) flow system for the substrate solution
(c) effect of column dimensions on the reaction rate
(d) pressure drop over the column.
Different maximal flow rates for complete hydrolysis were obtained for various substrates and the same conversion was observed whether upward or downward flow of
substrate solution was applied. Column dimensions had no effect on the reaction rate and
the pressure drop was proportional to flow rate and column length at a specific temperature. With these results in mind an enzyme reactor was designed. A flow diagram of
the complete system is shown in Fig. 3. The entire system is automatically controlled
and operated continuously.
Various amino acids were produced by passing a solution of acyl-D, L-amino acid
through a column containing 1000 l of DEAE-Sephadex-aminoacylase. With a 100%
deacylation of the L-isomer, a theoretical yield of beetween 6.4 and 21.5 tons per
month was calculated for various amino acids; the practical yield varied between 70
and 90% [121]. The column retained more than 60% of the original activity after
more than 30 days continuous operation at 50 °C. The half-life of the column was
estimated to be approximately 65 days. After 30 days operation the column was regenerated to its original activity simply by the addition of the amount of enzyme
activity corresponding to the lost activity. Another advantage of the system is that the
column can be regenerated without removing the support. The DEAE-Sephadex itself is
very stable and has been used for more than 5 years without any significant loss of
binding capacity [122].
An economic comparison between the conventional batch process with soluble enzyme
and the new continuous immobilized enzyme process shows that L-amino acids can be
produced at a lower cost by the new process [120]. In fact, the overall operating cost
of the immobilized enzyme process was about 60% of that of the process using soluble
enzyme. A number of factors contributed to this reduction of the production cost.
For example, the isolation of the product became simpler and the yield was increased.
Consequently, less substrate was required for the production of a unit amount of amino
acid. In addition to this, much less enzyme is needed and the labor cost is reduced
considerably due to automatic control of the process.
As an alternative carrier for aminoacylase, the use of porous glass beads has been in-

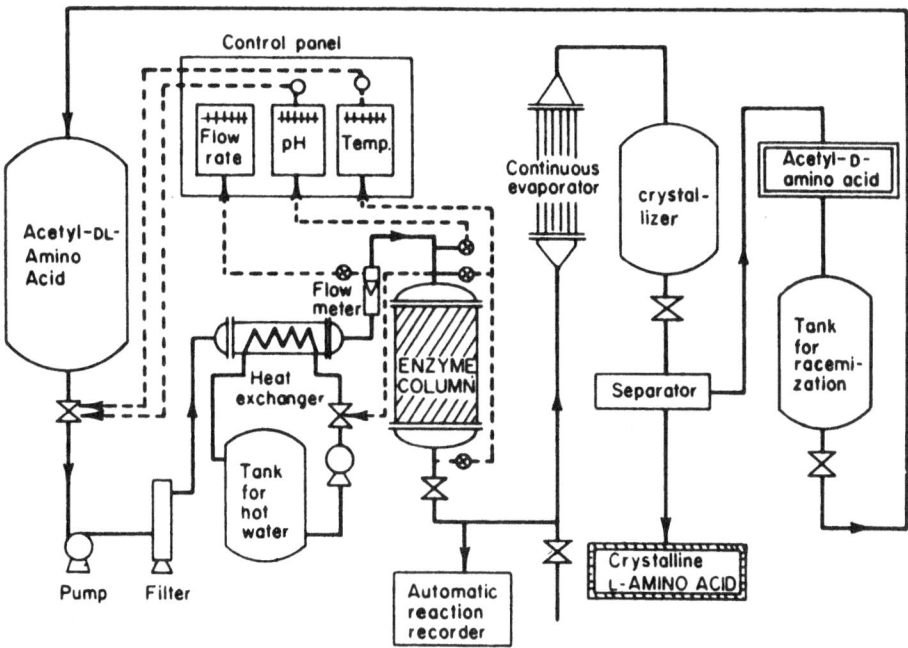

Fig. 3. Flow diagram for continuous production of L-amino acids by using immobilized amino-acylase (from [120])

vestigated [123]. Thus, aminoacylase from *Aspergillus sp.* was covalently bound to alkylamino- and arylamino derivized glass. Since the alkylamino derivative was the most active and stable preparation, it was chosen for more detailed studies.

The applicability of this preparation of aminoacylase was tested for the resolution of D,L-amino acids on an industrial scale over a long period of operation. A column containing an aminoacylase activity of 6.85 units per ml of glass beads was operated continuously at flow rates of SV = 5 and SV = 10 at 37 °C. At the lower flow rate the substrate, N-acetyl-D,L-methionine (0.1 M), was hydrolyzed to 48%, i.e. 96% of the L-isomer was deacetylated. Somewhat less hydrolysis of the substrate was observed at the higher flow rate and hydrolysis started to decline after approximately 40 days. On the other hand, the same degree of hydrolysis was still obtained after 60 days operation at the flow rate of SV = 5. This behavior indicates that a diffusional limitation existed. It was calculated that, at the lower flow rate, about 25 kg of L-methionine could be produced by continuous operation of 1 l of the immobilized enzyme for one month. This productivity is comparable to the one obtained with DEAE-Sephadex-amino-acylase.

Finally, aminoacylase has been immobilized by entrapment in cellulose triacetate fibers and used for the production of L-amino acids on a pilot plant scale [124]. A packed bed reactor, as part of a recycling process, was utilized at 45 °C for production of tryptophan, phenylalanine, methionine, and valine in good yields. For instance, after recycling a 0.2 M solution of acetyl-D,L-tryptophan for 5.5 h, L-tryptophan could be isolated in 93% yield. After 30 days the activity of the enzyme fibers was 85% of the initial activity.

Racemic mixtures of amino acids can also be resolved by other enzymatic reactions. There are several enzymes, proteases and peptidases, which specifically hydrolyze esters of the L-isomer of the amino acids. However, the corresponding ester of the D-isomer, is in most cases a strong inhibitor of these enzymes, and this approach for resolution of racemic mixtures is less attractive. Several proteases and peptidases have been immobilized and are used for other purposes as discussed in other sections (e.g. 3.2.2, 3.3.3, and 3.4.1).

In principle, it would be possible to resolve a racemic mixture by utilizing immobilized D-amino acid oxidase, which transforms the D-isomer to an α-keto acid leaving the L-isomer unreacted. In fact, immobilized amino acid oxidases are actually used in the production of α-keto acids as described in Section 4.4.1.

3.3.2 Biosynthetic Production of Amino Acids

Biosynthetic methods are widely used for the specific production of many amino acids. Such methods involving immobilized enzymes or microbial cells have been developed for the manufacturing of some amino acids on laboratory or industrial scale.

3.3.2.1 L-Aspartic Acid. Aspartic acid is widely used as a food additive and in medicine. It is industrially produced from fumaric acid and ammonia employing the enzyme aspartase as catalyst. This enzyme is produced by many microorganisms, such as *Escherichia coli*. Tanabe Seiyaku Co. has also investigated immobilization of this enzyme for development of an industrial process [125]. The intracellular enzyme from *E. coli* was partially purified. This preparation of aspartase contained substantial amounts of fumarase, which catalyzes an unwanted side reaction. This contaminating activity could be selectively destroyed by heat treatment. After immobilization of the heat treated enzyme preparation by ionic binding, covalent coupling, physical adsorption or entrapment, a comparison showed that the polyacrylamide entrapped aspartase was the most active.

When a solution of ammonium fumarate was passed through a column containing the gel-entrapped enzyme, asparatic acid was formed in high yield (95%). The half-life of the column was estimated to be 27 days at 37 °C, which was considered not to be satisfactory for industrial use.

Intracellular enzymes, as in this case, are generally unstable when extracted from the cell. Attempts to immobilize the whole microbe led to a more stable enzyme preparation, however, and eliminated extraction and purification [34, 126]. The reason for this marked increase in stability has been investigated in detail [127]. The binding of aspartase to cellular particles or membranes appears to play an important role in the stabilization of this enzyme.

Since 1973 the Tanabe company has utilized a system involving immobilized *E. coli* cells for the industrial production of L-aspartic acid. *E. coli* was selected from a number of microorganisms as having the highest aspartase activity. In this case, entrapment by polyacrylamide gel also gave the most active preparation. The activity of the cells increased during incubation in a solution at weakly alkaline pH and a high ionic strength. This activation (almost 10 times) is due to the autolysis of cells. A comparison of "activated" immobilized cells and the immobilized partially purified aspartase demonstrated that considerably higher activities were obtained with the comparatively simple immobilized whole cell system.

The reactor design used for the continuous production of L-aspartic acid was in principle the same as that developed for aminoacylase. A solution of 1 M ammonium fumarate containing 1 mM $MgCl_2$, pH 8.5 was passed through the column at a flow rate of $SV = 0.6$ at 37 °C. The amino acid could be isolated from the effluent in good yield (95%). The half-life of the aspartase column was estimated to be 120 days. The capacity of the plant is 13 tons of L-aspartic acid per month.

An economic analysis shows that the immobilized cell system is superior to the conventional batch process. The overall production cost was reduced to about 60% of that for the conventional process, with savings made on the cost of catalyst and labor. Recently the process has been improved further by utilizing a new immobilization technique [128]. Increased activity after immobilization and better operational stability was obtained when the bacterial cells were immobilized in a gel of carrageenan, a polysaccharide used as a food additive.

Aspartase has also been immobilized in the hollow centers of semipermeable fibers [129]. A solution of aspartase from *E. coli* was introduced into viscose fibers under pressure. The fibers were knitted into a sheet and the ends were sealed in order to entrap the enzyme. The knitted material was used in a batch reactor by dipping the sheet in a 10% solution of ammonium fumerate at 37 °C for 8 h. In several consecutive experiments, the yield of L-aspartic acid exceeded 80%.

For the continuous utilization of such knitted sheets containing immobilized enzyme, different reactor designs are possible. The sheets can be staggered in a column or used in a laminar arrangement. A third possibility is to coil the sheet into a spiral in analogy with a reactor type developed for enzyme-collagen membranes [23].

3.3.2.2 L-Glutamic Acid. Most glutamic acid is produced by fermentation utilizing the biotin-requiring *Corynebacterium glutamicum,* which can accumulate as much as 75 mg of glutamic acid per ml of culture medium. The acid can also be obtained by extraction of vegetable proteins after hydrolysis and the DL-form by chemical synthesis. An interesting approach to produce glutamic acid with polyacrylamide entrapped cells of *C. glutamicum* is based on this fermentative process with glucose as only carbon source [37]. After 120 h incubation, the immobilized cells had produced 13-14 mg per ml of glutamic acid. After 3 month storage the same gel preparation could produce the same amount of glutamic acid during a 150 h incubation. Even though no evaluation of the stability and productivity of the immobilized cell preparation was carried out the results are noteworthy because they demonstrate the feasibility of utilizing a complex multienzyme pathway of an intact immobilized microorganism.

3.3.2.3 L-Tryptophan. L-Tryptophan can effectively be produced by a biosynthetic method involving tryptophanase, which catalyzes α,β-elimination and β-replacement reactions. The former reaction, which is reversible, can be used for the synthesis of tryptophan from indole, pyruvate and ammonia. The enzyme from *E. coli*, which requires pyridoxal 5-phosphate for activity and consists of four identical subunits, has been immobilized on agarose by different methods [130]. The most efficient immobilization (81%) and the highest relative activity (60%) was achieved when the apoenzyme was adsorbed to Sepharose-bound pyridoxal phosphate. The affinity binding was stabilized by reduction of the Schiff base with $NaBH_4$, and the enzyme was reconstituted by washing with pyridoxal phosphate. A gradual decrease of activity was observed during repeated use in the absence of coenzyme.

L-Tryptophan or its analogue 5-hydroxy-L-tryptophan was synthesized in good yields by a continuous flow method or by a repeated batch process. With a column containing tryptophanase-Sepharose (10 ml) about 90% of the indole (0.8 mM) was converted to tryptophan at an appropriate flow rate (SV = 5) at 37 °C. At this temperature the immobilized enzyme lost less than 5% of its original activity during 5 days. Under the same conditions the soluble enzyme lost about 90% of its acitivity. A remarkable stability is thus observed for the immobilized enzyme and its seems to be possible to use this preparation as catalyst in longterm reactions.

A partially purified tryptophanase from *E. coli* has been immobilized by entrapment in cellulose triacetate fibers for the production of L-tryptophan from indole and D,L-serine in batch and continuous processes [131, 132]. At high indole concentrations, the enzyme activity was reduced due to adsorption of indole to the fibers leading to an increased diffusional barrier. In the batch process, the initial concentration of indole was 0.5 g per l. Solid indole was added at 1 h intervals in portions equivalent to that transformed to tryptophan. After 8 h the reaction was stopped and the tryptophan concentration in the mixture was then 4.4 g per l.

In the continuous feed recycle reactor the substrate solution was pumped through the column (6 × 60 cm) at a flow rate of 25 l per h and with a feed/outlet of 0.8 l per h [132]. The maximum applicable indole concentration was determined to be 350 mg per l in the feed solution and the concentration in the outlet maintained constant at 8 and 510 mg per l for indole and tryptophan, respectively. The reactor could be operated for 20 days without any loss of tryptophan productivity.

The authors concluded that the process may be competitive with fermentative processes. Some of the advantages they pointed out included a relatively stable enzyme preparation, the possible re-use of D-serine after racemization, and easy isolation of the product.

Also, whole *E. coli* cells containing tryptophanase have been immobilized by entrapment with polyacrylamide [133, 134]. Tryptophan (5-hydroxytryptophan) was produced from indole (5-hydroxyindole) and serine or ammonium pyruvate. The yield of tryptophan and 5-hydroxytryptophan was 86% and 72%, respectively, after incubation at 30 °C for 24 h.

3.3.2.4 L-Tyrosine. In analogy with tryptophanase, β-tyrosinase catalyzes α,β-elimination and β-replacement reactions. Thus, by employing the reversible elimination reaction, tyrosine can be synthesized from phenol and ammonium pyruvate or serine. Also L-DOPA, an important drug for treatment of Parkinson's disease, can be produced by substitution of phenol by pyrocatechol in the reaction mixture.

The enzyme from *Escherichia intermedia*, which requires pyridoxal phosphate for activity, was immobilized by covalent linkage to BrCN-activated Sepharose [130]. In this case adsorption to the immobilized coenzyme is less attractive because the enzyme is a dimer and therefore half of the active sites would be required for immobilization, resulting in low activity. The preparation of immobilized tyrosinase was used for the continuous production of L-tyrosine and L-DOPA. During 5 days of operation at 25 °C, pH 8.0, the column lost 30% of its original activity.

3.3.2.5 L-Lysine. Many systems involving immobilized enzymes or microbial cells have been developed for the synthesis of L-lysine, an essential amino acid, which is currently made by a fermentative process.

One of these is a process involving a two-enzyme system, consisting of L-α-amino-ε-caprolactam hydrolase (from *Cryptococcus laurentii*) and α-amino-ε-caprolactam racemase (from *Achromobacter cyclocraster*), as outlined in Fig. 4 [135]. The starting material, D,L-α-amino-ε-caprolactam, which is easily synthesized from cyclohexane, a by-product of nylon synthesis, is asymmetrically hydrolyzed by the first enzyme. The second enzyme converts the D-isomer to the racemic D,L-form. The two enzymes were immobilized through ionic binding to DEAE-Sephadex and used in a column at 42 °C. L-Lysine was synthesized in high yield (95%) from D,L- or D-α-amino-ε-caprolactam.

In another process polyacrylamide entrapped diaminopimelate decarboxylase was used to produce L-lysine from meso-2,6-diaminopimelic acid [136]. Before immobilization the enzyme was purified 20-30 fold from a crude extract of *E. coli*. In a batch process the product was formed in approximately 70% yield after 2 h incubation at 40 °C. This process was also extended with an additional enzyme, diaminopimelate racemase, which was entrapped separately in polyacrylamide after partial purification from an extract of *Microbacterium ammoniaphilum* [137]. Threo-2,6-diaminopimelic acid was converted to 46% of the meso-isomer by incubation at 40 °C for 2 h in a batch process with immobilized racemase. The reaction mixture was then incubated at 50 °C for 2 h with the immobilized decarboxylase. The yield of L-lysine, which was relatively low, may be improved considerably by co-immobilization of the two enzymes.

3.3.2.6 L-Alanine. The production of L-alanine from aspartic acid with immobilized cells containing aspartate β-decarboxylase has been described [138, 139]. Cells of *Pseudomonas tagnei* or *Pseudomonas dacunhae* were immobilized by entrapment

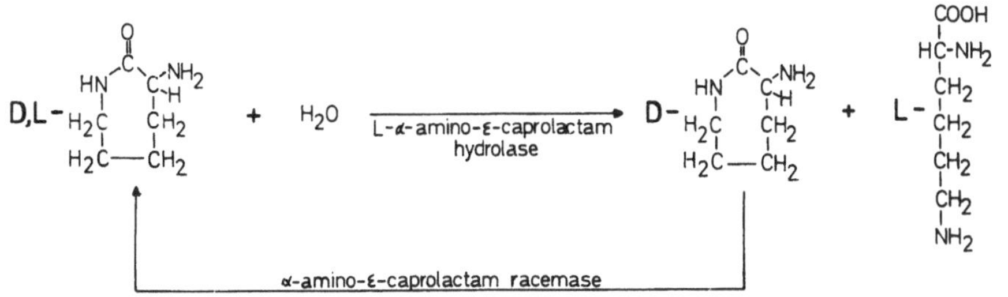

Fig. 4. Production of L-lysine from D, L-α-amino-ε-caprolactam with a two-enzyme system

in polyacrylamide and packed in columns. D,L- or L-aspartic acid (1 M) was continuously passed through these columns at a flow rate of SV = 0.16 at 37 °C and L-alanine was isolated from the effluent in good yield (82% and 90% respectively for D,L- and L-aspartic acid as substrate). When D,L-aspartic acid was used D-aspartic acid could also be isolated (92%).

3.3.2.7 L-Citrulline. L-Citrulline can be produced from L-arginine by the action of L-arginine deiminase, which is present in various microorganisms. This enzyme activity is often associated by ornithine transcarbamylase, which converts citrulline to ornithine. The bacterium *Pseudomonas putida* having high activity of deiminase and no activity of transcarbamylase has been used in a batch process for the production of L-citrulline. To overcome some disadvantages of this process the bacterium was immobilized with polyacrylamide and used for the continuous production of the amino acid [140]. After immobilization 56% of the enzyme activity was recovered. The decrease of activity was believed to be partly due to restricted diffusion of substrate and product in the gel lattice. The immobilized cells did not, as did the intact cells, require a surfactant for activity. Obviously, a diffusion barrier for arginine and/or citrulline had been damaged by the immobilization.

When a solution of L-arginine (0.5 M) was passed through a column containing the entrapped cells at a flow rate of SV = 0.26 at 37 °C, the substrate was completely converted to L-citrulline. The Product was readily isolated from the effluent. The operational stability was very high, i.e. the half-life of the immobilized cells was 140 days at 37 °C [141]. This continuous production of L-citrulline was considered to be a more useful technique for the commercial production of this amino acid than the methods using microbial processes.

3.3.3 Hydrolysis of Proteins

Production of amino acids from protein hydrolysates obtained by the action of proteases and peptidases is in principle feasible but problems in isolating individual amino acids make this procedure less attractive. On the other hand this approach is of interest when proteins need to be partially or completely hydrolyzed for solubilization or increased digestability. The use of immobilized enzymes for this purpose is very attractive since many proteases and peptidases are expensive and only a few of these enzymes come from organisms on the GRAS (generally recognized as safe) list. Furthermore, a process based on immobilized enzymes offers the standard advantages like continuous operation, process control, etc.

β-Lactoglobulin was hydrolyzed almost completely to free amino acids by a combination of pronase and leucine aminopeptidase separately immobilized on porous glass [142]. Treatment with only the immobilized pronase resulted in 70% hydrolysis. Another substrate, casein, at a concentration of 4-5% solids, was hydrolyzed to 80% by immobilized peptidases and pronase [143].

These studies indicate that hydrolysis of proteins to free amino acids is possible but more active preparations are required before the process can be developed further. In order to achieve this, new enzymes and combinations of enzymes must be examined.

3.4 Miscellaneous Applications in Food Industry

3.4.1 Beer Industry

Beer that has not been chillproofed and has been stored for a few weeks at room temperature can develop a haze upon chilling. This haze has been determined to be complexes of tannins, proteins, and carbohydrates. Several methods are used to eliminate the chill haze problem. Different adsorbents can be used to remove haze components more or less specifically. A more sophisticated technique, which has been used for more than half a century, is to hydrolyze the haze protein into polypeptides and amino acids with a proteolytic enzyme. However, excessive hydrolysis can affect foaming and beer quality. Papain has been found to be quite suitable since its action is reasonably selective and can be controlled under manufacturing conditions. Thus, chillproofing is carried out by addition of a solution of papain to the beer and after standing for 3—4 days at 4 °C the beer is pasteurized, whereupon the papain is inactivated.

Immobilized papain appears to be very suitable for the chillproofing operation. The processing could be carried out continuously by passing the beer through a column containing a highly active immobilized papain preparation. Consequently, the time required for the process would be reduced considerably as well as the size of the needed equipment. An additional advantage, in view of possible restrictions on the addition of enzymes to foodstuffs in the future, would be that a product free of external protein is obtained.

Papain has been immobilized to various supports for use in chillproofing of beer [144–149]. Papain-collagen membranes were used in a continuous reactor for 5 months with 56% of the initial activity remaining [144]. With appropriate flow rate a product of acceptable haze stability was obtained. By increasing the specific activity of the papain used and increasing also the loading of the column, the performance of the reactor could be improved.

A preparation of papain covalently bound to a hydroxyalkyl metacrylate gel was tested in both batch and column reactors [145]. The effects of temperature, enzyme concentration, and contact time were investigated. Even though an increased thermostability of the immobilized papain was observed it was concluded that for practical reasons the reaction should be carried out at 10 °C. The lower reaction rate at this temperature can be compensated for by either a longer reaction time or addition of more enzyme preparation in a batch operation and by a lower flow rate in a continuous column operation. Low concentrations of immobilized enzyme (less than 1 g per l beer) reduced the amount of substances precipitable by ammonium sulfate and higher concentrations (more than 1 g per l beer) resulted in a more extensive hydrolysis of proteins into low molecular weight polypeptides and amino acids. Since this extensive hydrolysis of proteins is undesired and not necessary, careful optimization of the reaction conditions should be carried out.

An interesting reactor design involving immobilization of papain on cotton yarn was obtained in the following way [146]. Cotton yarn was winded on a roll of stainless steel screen to give a cartridge which was placed in a stainless steel holder. The cellulose was activated by recycling a $HClO_4$-solution through the reactor and, after washing with water, the enzyme was immobilized simply by recycling a solution containing papain through the column. The reactor was then used to chillproof beer at 5 °C. A single pass of beer through the reactor was not sufficient but after recycling a good haze value was obtained.

There are a number of other processes in the brewing industry where immobilized enzymes have a potential use [150]. By adapting and perhaps modifying processes developed for the sugar industry many glycolytic enzymes might be useful in beer production. Immobilized microbial α-amylase could be added to assist or even replace the natural amylases of malt. Production of a highly fermentable sugar wort could be achieved by the action of immobilized α- and β-amylases and glucoamylase. In addition to this, immobilized β-glucanase might be useful in adjustment of beer viscosity since β-glucan contributes to the viscosity of beer. All these potential applications are only possibilities of future developments.

Immobilized glucoamylase has, however, been used for the hydrolysis of dextrins which are present in beer [151]. If the treated beer contained residual yeast cells the produced glucose could be utilized in a second process. Fully fermented yeastfree beers were treated with the insoluble glucoamylase at 5 °C for 3 days to produce glucose at 0.6%. These beers remained clear and were of satisfactory flavor.

3.4.2 Juice and Wine Industries

The treatment of beer with immobilized proteinases for the purpose of stabilizing it against haze formation has just been described. Similar problems may occur in the manufacturing of other beverages such as sake, fruit juices, and wine. The utilization of immobilized enzymes to solve these problems has been considered advantageous because no external protein would remain in the final product. Consequently, any enzyme source can be used including organisms which are not on the GRAS list. Usually protein turbidity appears in pasteurized and stored sake, a wine made from rice. Clarification of sake can be carried out by proteolytic hydrolysis of proteins causing the turbidity. The protein fragments from such a treatment will precipitate upon storage and thus leave a clear sake. A thermophilic protease from *Penicillium dupontii* immobilized on BrCN-activated Sepharose was tested for the continuous clarification of sake [152]. Immobilization of the enzyme increased its thermostability and it was possible to combine the pasteurization step and the enzymatic treatment by passing fresh sake through a reactor loaded with the immobilized protease at 65 °C at a mean holding time of 3 min. Relatively high enzyme activities were obtained after immobilization and also good stability was observed. However, for practical application a less expensive carrier and a simpler immobilization technique must be employed. Immobilized proteases have also been used for clarification and stabilization of grape juice and wine [153, 154]. Pepsin and acid proteinases isolated from the fungi *Aspergillus awamori* and *Aspergillus oryzae* immobilized on bromoacetyl cellulose and porous carriers, such as silica gel und glass, proved most promising for application in juice and wine production [153]. The half-lifes of the preparations were 7–8 days under operational conditions in a column. Somewhat different enzyme stability was observed when juice and wine were treated, which was in part due to the denaturing action of ethanol and in part to phenolic and other compounds present in the wine in larger amounts than in juice. Furthermore, it has been pointed out that Fe^{2+} and Fe^{3+} ions present in grape juice inhibit the immobilized acid proteinase [154]. In a technical process, which was elaborated for industrial scale, these ions were removed by an ion-exchanger prior to the proteolytic treatment. In this process the protein content of the juice was decreased by more than three fold. The stability of the product, which was crystal clear and had a good taste, was increased from 7–10 months to 13–14 months depending on the grapes used.

3.4.3 Production of Nucleotides

To enhance the flavor of meat, monosodiumglutamate is added and this flavor enhancement is increased considerably by addition of an equimolar mixture of the nucleotides inosinic acid (IMP) and guanylic acid (GMP). Thus, the flavor enhancement property of the glutamate is increased by a factor of five when it is fortified with 4% of a mixture of IMP and GMP.

In Japan large amounts, estimated at 2000–3000 t per year in 1973, of the nucleotide mixture is produced for this purpose. One of the main commercial processes for the production of IMP and GMP is based on enzymatic hydrolysis of RNA which is isolated from yeast [155]. Initially, the RNA is hydrolyzed by 5'-phosphodiesterase to its constituent ribonucleotides and in a second step AMP is converted to IMP by the action of 5'-AMP deaminase. Subsequently, the nucleotides of the treated hydrolysate are separated by ion-exchange chromatography.

In order to evaluate a continuous process for the production of IMP and GMP, phosphodiesterase from *Penicillium* sp. and AMP deaminase from *Aspergillus* sp. were separately immobilized by covalent coupling to porous ceramic [156]. Since encouraging results on enzyme stability and activity were obtained, a large scale operation was evaluated. To compare this new process based on immobilized enzymes and the conventional batch process utilizing soluble enzymes, the following process parameters were assumed:

capacity:	1000 kg RNA per day
feed solution:	4% and 10% RNA for the continuous and the batch processes, respectively
conversion rate:	80% hydrolysis of RNA; 100% deamination of AMP
enzyme utilization:	immobilized enzyme replaced every 30 days; soluble enzyme not recovered
yield:	IMP 158 kg and GMP 168 kg per day.

Whereas the batch process required a reactor volume of 15 m^3 the volume of the immobilized enzyme reactors were only 0.6 and 0.3 m^3 for the esterase and the deaminase, respectively. The total cost of the catalyst (enzyme and carrier) for the continuous process was found to be about 25% of the cost of the batch process. Although the enzyme cost constitutes a rather small part of the variable cost, the merits of the immobilized enzyme process come from decreasing fixed costs, such as labor and construction costs.

4 Applications in the Pharmaceutical Industry

In the pharmaceutical industry, many chemical transformation are carried out by biocatalysts, such as purified enzymes or whole microbial cells. Many of these processes
are well suited for the use of immobilized enzyme technology. This applicability is
emphasized by the fact that drugs are produced on a scale adequate for procedures
involving immobilized biocatalysts. In fact, an increasing number of pharmaceuticals
are produced by the employment of immobilized enzymes as will be described in the
following sections.

4.1 Penicillins

6-Aminopenicillanic acid (6-APA) is an important intermediate compound for the
production of semisynthetic penicillins. The antibiotic activity of the penicillin molecule is influenced by the side-chain substituent at the 6-amino group. Thus, various
penicillins are synthesized from 6-APA by substitution at this position. 6-APA can be
produced on a large scale from penicillin G (benzyl penicillin) or penicillin V (phenoxy-
methyl penicillin), both of which are produced by fermentation, by removal of the
side chain according to Fig. 5.
For this hydrolysis two processes, one chemical and one enzymatic, are available. The
enzymatic process is widely applied and during recent years the conventional process
based on soluble enzyme has been converted to an immobilized enzyme process.
Several pharmaceutical companies, e.g. Beecham, Bayer, Squibb, and Astra, have
developed their own immobilized enzyme processes, which now are operating on an
industrial scale.
Many preparations of immobilized penicillin acylase have been described in the patent
literature and it is not stated which of several preparations actually are being used in
the industrial processes. Furthermore, it is difficult to obtain further information from
the pharmaceutical industry. An exception is the system developed by Astra (Sweden)
which has been described in some detail [157, 158]. Snam Progetti (Italy) has developed
a process involving entrapment of penicillin acylase in cellulose triacetate fibers, which
also has been described to some extent [159, 160]. Tanabe Seiyaku Co. (Japan) has a
process for continuous production of 6-APA by immobilized whole cells, which has
been published [38]. These three systems will be reviewed and other systems described
in the patent literature will be briefly discussed.

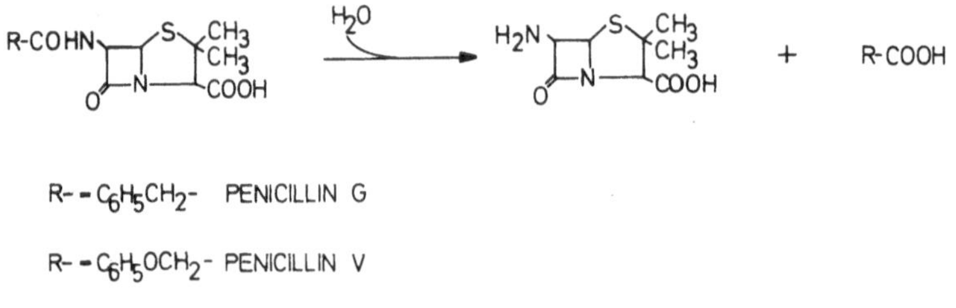

Fig. 5. Formation of 6-aminopenicillanic acid by hydrolysis of penicillins

The enzyme used for the hydrolysis, penicillin acylase (or penicillin amidase), can be obtained from various microorganisms. A batch or recirculation reactor is preferred since acid is formed during the reaction, which has to be neutralized.

In the Astra process penicillin acylase purified from *Escherichia coli* was immobilized on BrCN-activated Sephadex 200 [157, 158]. After coupling, 48% of the enzyme activity was recovered on the polymer, which had an activity of about 225 units per g of wet polymer. The enzyme preparation has been used in a batch process since 1973 for the industrial production of 6-APA. In a batch 100 kg of penicillin G was hydrolyzed by 16.5 kg of wet Sephadex-enzyme complex (corresponding to 3.7×10^6 units) at 35 °C, pH 7.8. The immobilized enzyme could be used for more than one hundred batches without addition of fresh enzyme, provided that the operations were carefully performed. Special care must be taken in the recovery (filter press) and recharging of the enzyme to avoid losses of activity. To eliminate these problems a recirculation process has later been introduced in the factory. This reactor design had some additional advantages such as higher production capacity and lower costs. Compared to the old method using whole microbial cells in suspension for the cleavage of penicillin G this new process has resulted in higher yields of a purer product, easier handling of the catalyst and better economy. In the recirculation process, operated at 37 °C, pH 7.8, 6-APA of 98% purity was obtained in 90% yield.

It was concluded that no leakage of protein from the support occured and therefore 6-APA and semisynthetic penicillins of hypoallergenic quality could be produced. Normally proteins have to be removed from the 6-APA in a special purification step. Presence of such proteins in the final product can result in serious allergenic reactions during penicillin therapy. It should be pointed out that specially purified penicillin G was used in the process.

A somewhat different approach has been taken by Snam Progetti which has developed a process for the production of 6-APA from a crude preparation of penicillin G [160]. Normally, the penicillin G is extracted from the fermentation broth and crystallized by azeotropic destillation. During the crystallization step 8-10% of the penicillin G is lost. A process has been developed to eliminate the crystallization step and the accompanying loss of penicillin.

Partially purified penicillin acylase from *E. coli*, entrapped in cellulose triacetate fibers, was used in a column as a part of a recirculation batch reactor with continuous titration of liberated acid. Initially, when a crude substrate solution was used, a drastic decrease in enzyme activity was observed, which was caused by a compound extracted from the microbial broth together with the penicillin G. A method to remove this compound was developed and the crude penicillin G could subsequently be hydrolyzed without any loss of activity. A 6% substrate solution was hydrolyzed to at least 90% during 1.5–2 h reaction time. The half-life of the column was about 2 months at 37 °C.

An economical comparison between the chemical, the conventional enzyme and the new integrated immobilized enzyme process for a production capacity of 40000 kg of 6-APA per year has been carried out. The total cost for the conventional enzyme process turned out to be 7-8% lower than for the chemical process, while the total cost for the new process was about 20% lower. Many factors contributed to this reduction of cost such as lower capital investments due to a simpler processing scheme and lower cost for utilities. For instance, while the enzymatic hydrolysis is carried out at ambient temperatures, the chemical process requires a reaction temperature of $-70\,^\circ$C which is maintained with expensive liquid nitrogen. Furthermore, a higher yield of product is obtained in the integrated process and recovery and re-use of phenylacetic acid is possible in the new process. It should, however, be pointed out that the integrated process is practical only when the fermentation plant and the 6-APA production plant are situated at the same location.

Tanabe Seiyaku Co. has chosen still another approach. In line with other developments (see Sections 3.3.2.1 and 4.4.3), whole microbial cells (*E. coli*) were immobilized by entrapment with polyacrylamide gel to produce 6-APA [38]. This was done in order to eliminate the purification and increase the stability of the enzyme. Cells having high penicillin acylase activity but also some penicillinase activity were selected. The former activity was about 10 times higher than the latter and by optimizing the reaction time, a relatively small amount of 6-APA was decomposed by the penicillinase. By passing a solution of penicillin G (50 mM) through a column containing the immobilized cells at a flow rate of SV = 0.24 at $40\,^\circ$C, pure 6-APA could be produced in 78% yield in a subsequent crystallization step. The half-life of the column under continuous operation was estimated to be 42 days at $30\,^\circ$C and 17 days at $40\,^\circ$C. It is interesting to note that even though an enzyme decomposing the product is contained in the cells, the immobilized cells can be used as biocatalyst when the difference in reaction rate between desirable and undesirable reactions is significant.

Beecham (England) has coupled penicillin acylase from the same organism covalently to a polymethacrylate resin with glutaraldehyde [161]. The preparation was used to hydrolyze 6% penicillin G in 15 successive batches. The same enzyme has also been coupled to a water-soluble polymer, a copolymer of sucrose and epichlorohydrin [162]. The molecular weights of the enzyme complexes were in the range of 125,000 to 250,000 and they were used in an ultra-filtration reactor to hydrolyze penicillin G or penicillin V. 6-APA could be obtained in about 90% yield in a series of batches.

Bayer (Germany) has also employed penicillin acylase from *E. coli*. The enzyme was coupled to a copolymer of acrylamide, N,N'-methylenebisacrylamide, and maleic acid anhydride with a method resulting in covalent binding of 98% of the enzyme added to the polymer [163]. In this case 6-APA was produced from penicillin G in 86-89.5% yield at $38\,^\circ$C, pH 7.8 [164].

Penicillin acylase from *Bovista plumbea* was immobilized by entrapment in cellulose acetate fibers for the production of 6-APA from penicillin V by Biochemia GmbH (Germany) [165]. The fibers were packed into a column and a solution of penicillin was recycled at $32\,^\circ$C, pH 7.5. The liberated acid was neutralized with 10% ammonia. After 3 h, 97% of the substrate had been cleaved into 6-APA and phenoxyacetic acid. After crystallization the yield was 91.5%.

Squibb (USA) has produced 6-APA with a preparation of penicillin acylase from *Bacillus megaterium* adsorbed onto bentonite [166, 167]. The extracellular enzyme was immobilized simply by adding bentonite together with a filtration aid to the fermentation broth after removal of the microbial cells. Penicillin G was effectively hydrolyzed in either batch or continuous reactors. An interesting finding was that the adsorbed enzyme actually exhibited a higher productivity than the soluble enzyme.

This effect was ascribed to reduced product inhibition upon immobilization [167]. Finally, a process for production of 6-APA, involving immobilized acylase from *E. coli*, has for some time been operating in the USSR [168]. The enzyme entrapped in polyacrylamide gel, is used for hydrolysis of penicillin G in a batch reactor [169, 170].

In principle, the acylation of 6-APA to various penicillins may be carried out enzymatically. There are only a few reports on such enzymatic acylations of 6-APA, because the reaction has little commercial utility. The reasons for this are a relatively low yield of acylated product and the availability of simple chemical methods of acylation. Penicillin acylase from *E. coli*, entrapped in cellulose triacetate fibers, was, however, used in a continuous flow reactor to produce ampicillin and amoxycillin from 6-APA and D-phenylglycine methyl ester and D-p-hydroxyphenylglycine ethyl ester, respectively [171]. In all experiments the conversion was less than 50% and thus not attractive for large scale production.

Toyo Jozo Co. Ltd. (Japan) has developed a process involving cells of *Bacillus megaterium* or *Achromobacter sp.* adsorbed to DEAE-cellulose for production of ampicillin [172]. The penicillin was, for instance, produced in 54% yield by passing a solution of 6-APA (0.3%) and D-phenylglycine methyl ester (0.9%) through a column containing the immobilized cells of *Achromobacter sp.* at a flow rate of SV = 0.5.

Otsuka Pharmaceutical Co. Ltd. (Japan) has used succinoylated penicillin acylase adsorbed on DEAE-Sephadex to synthesize ampicillin [173]. To minimize enzyme leakage, the enzyme was succinoylated to indroduce a number of carboxylic groups thereby increasing the binding force between enzyme and solid support. The enzyme-carrier complex was used in a column at 37 °C and the yield of ampicillin was 67%, which after ten runs had decreased to 58%.

4.2 Cephalosporins

Cephalosporin antibiotics can be synthesized from penicillins or made by direct fermentation. *Cephalosporium acremonium* produces cephalosporin C (I in Fig. 6), which contains the 7-aminocephalosporanic acid nucleus (7-ACA) (II in Fig. 6), and the side chain α-aminoadipic acid. Other cephalosporins containing the 7-ACA nucleus have to be produced from cephalosporin C by deacylation and subsequent acylation to introduce the desired side chain. Deacylation is accomplished chemically with reagents such as PCl_3 or nitrosyl chloride [174]. The acylation can be carried out either chemically or enzymatically.

Cephalosporins can also be produced chemically from penicillins by expanding the 5-membered thiazolidine ring of penicillin to the 6-membered dihydrothiazine ring of cephalosporins. In this case the side chain can be removed chemically or enzymatically to obtain the nucleus 7-aminodesacetoxycephalosporanic acid (7-ADCA III in Fig. 6), which is an important intermediate in the synthesis of other cephalosporins. Chemical as well as enzymatic methods are available for the acylation of 7-ADCA. A great advantage of the enzymatic over the chemical acylation is that no blocking of reactive groups is necessary.

Toyo Jozo Co. Ltd. (Japan) has developed and patented several immobilized enzyme systems for the production of various cephalosporin derivatives [175–180]. For the

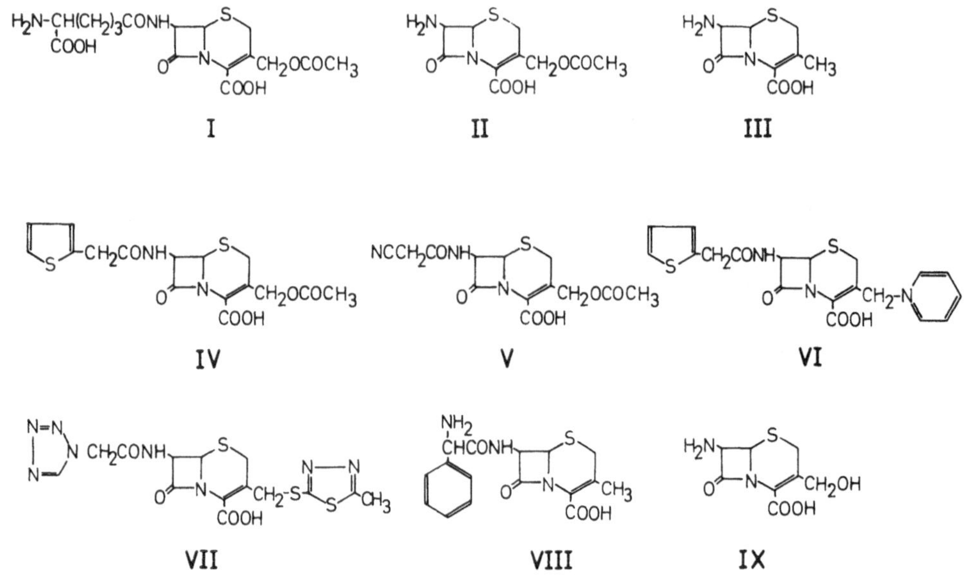

Fig. 6. Structural formulas of cephalosporins derivatives. I: cephalosporin C; II: 7-aminocephalosporanic acid (7-ACA); III: 7-aminodesacetoxycephalosporanic acid (7-ADCA); IV: cephalothin; V: cephacetrile; VI: cephalolysine; VII: cephazoline; VIII: cephalexin; IX: deacetyl-7-aminocephalosporanic acid

production of 7-ADCA from phenylacetyl-7-ADCA a deacylating enzyme of *Bacillus megaterium* was utilized [175]. The extracellular enzyme was adsorbed onto celite and packed in a column. Crystalline 7-ADCA (1.13 kg, corresponding to 85% yield) was produced during four days continuous operation of a column (10 l) at a flow rate of SV = 0.5 at 37 °C. No decrease in productivity was observed after 15 days operation. A deacylating enzyme from the culture filtrate of *Proteus rattgeri* adsorbed onto celite in a similar manner was also utilized for the production of 7-ADCA [176]. The product was of 93% purity and formed in 90% yield.

The acylase from *B. megaterium* adsorbed on celite has also been employed to carry out the reverse reaction, i.e. the acylation of 7-ACA or 7-ADCA. The commercially important cephalothin (IV in Fig. 6) was synthesized from 7-ACA and 2-thiophene acetic acid by continuously passing the substrate solution through the acylase column [177]. Crystalline cephalothin (1.34 kg, corresponding to 85% yield) was recovered from the effluent. Cephacetrile (V in Fig. 6) was produced in good yield by reacting a solution containing 7-ACA and cyanoacetate with the adsorbed enzyme for 90 min at 37 °C in a batch reactor [178]. Cephalolysine (VI in Fig. 6) [179] and cephazoline (VII in Fig. 6) [180] were made in a similar way in yields of about 75%.

All the above systems were developed by Toyo Jozo Co. Ltd., but other pharmaceutical firms have utilized immobilized enzymes for production of cephalosporin derivatives. Thus, Banyu Pharmaceutical Co. Ltd. (Japan) used a deacylase from *Arthrobacter viscosus* adsorbed on a mixture of calcium phosphate and Dicalite [181]. 7-ADCA was produced in a batch system at 37 °C from phenylacetyl-7-ADCA and after 3 h reaction time the product was isolated in 68% yield.

For the deacylation of the same substrate to 7-ADCA, Bayer AG (Germany) used a penicillin acylase immobilized by covalent linkage to soluble starch [182]. The resulting soluble starch-enzyme complex was occluded in an insoluble carrier, polyacrylamide gel. By such an immobilization technique much less leakage of enzyme is to be expected than if the enzyme itself was entrapped in the polyacrylamide gel. Furthermore, a gel with a looser structure can be used permitting substrate and product to diffuse easier into and out of the interior of the polymer. After 5 h reaction time at 37 °C, 7-ADCA of 97.5% purity could be isolated in 82% yield.

The fiber-entrapped penicillin acylase used by Snam Progetti for production of 6-APA [159] also catalyzes the synthesis of clinically important cephalosporins, such as cephalexin (VIII in Fig. 6) [171]. With a four molar excess of the side chain, i.e. D-phenylglycine methyl ester, 75% of the 7-ADCA was converted to cephalexin in one hour at 25 °C. If longer reaction times were employed the yield of product decreased. This was probably due to the acylase catalyzing the reverse reaction, i.e. deacylation of cephalexin.

The biological activity of cephalosporins does not only depend on the side chain at the C-7 position but also on the substituent at the C-3 position. Some modifications at the latter position can only be made after deacetylation of the molecule. Enzymatic hydrolysis is utilized to prepare deacetyl-7-ACA (IX in Fig. 6) in high yields. The cephalosporin acetylesterase of *Bacillus subtilis* has been shown to be very stable in solution and therefore it has been immobilized by containment in an ultrafiltration reactor [183]. The enzyme was reused 20 times over a period of 11 days to deacetylate 7-ACA of various concentrations. During this period 52% of the initial activity was lost but this corresponded well with the loss of 51% of the protein. If such losses of protein can be prevented the enzyme may be used for an extended period of time.

Ciba-Geigy Ltd. (Switzerland) has used an acetylesterase from *B. subtilis* with broad substrate specificity for deacetylation of cephalosporins [184]. The intracellular enzyme was relatively costly and therefore immobilization of the enzyme was conducted to enable reuse. Brick powder was chosen as carrier because it was inexpensive and had a relatively large internal surface area, as well as high dimensional stability, which allowed operation at high flow rates. The β-lactam ring of the cephalosporin molecule was subjected to pH-dependent non-enzymatic hydrolysis. High activity of the carrier-enzyme complex was therfore needed to keep the reaction time short to minimize non-enzymatic side reactions. The operation of a continuous packed bed reactor was hindered to some extent by the requirement of high buffering capacity to neutralize the generated acid. Substrate solutions about 20 mM were, however, hydrolyzed successfully when the buffer concentration was at least twice that of the substrate. Under these conditions the enzyme was stable for one month. Much lower stability was observed when a recirculation batch reactor operated under pH-stat conditions at high flow rates was used. The choice of reactor design is in this case partly influenced by practical problems associated with the separation of product and buffer salts.

4.3 Steroids

Many steroid transformations are very difficult to carry out chemically and therefore biocatalytic conversion is often the only practical way to obtain a particular steroid derivative. Even though in most instances only one enzyme is involved, large scale steroid transformations do not employ cell-free enzyme preparations. Since many of the enzymes utilized for steroid transformations require cofactors, which must be continuously regenerated during the reaction, whole microbial cells, containing systems for such regenerations, may be successfully employed. A great probability for undesired side reactions is prevalent, since the microorganism might contain other steroid transforming enzymes and careful selection of the organism to be used for a particular transformation is therefore important for a satisfactory result. The utilization of immobilized biocatalyst for steroid transformations has been reported but it appears that commercialization of such processes is not going to take place in the near future.

The synthesis of hydrocortisone and prednisolone from Reichstein's compound S has been studied with biocatalysts immobilized by entrapment with polyacrylamide [185, 186]. This interest can be explained by the increasing demand of these steroids for use as contraceptives and anti-inflammatory agents. For the first reaction, 11-β-hydroxylation, mycelia of *Curvularia lunata* and for the second reaction, Δ^{1-2}-dehydrogenation, whole cells of *Corynebacterium simplex* were employed. When high microbe densities were used during the immobilization step relatively good recoveries of activity were observed. Optimum substrate concentrations were determined to be 0.6 mM and 1 mM for the hydroxylation and dehydrogenation reactions, respectively. At higher concentrations substrate inhibition occured. It is important that the substrate conversion not only is specific, but also complete since otherwise tedious separation procedures have to be employed to obtain a pure product. In fact, the immobilized microorganisms transformed the substrate specifically and completely after appropriate reaction time. One of the most interesting and very important observations was, however, that when the activity of the immobilized microbes decreased to an impractically low level, the cells could be reactivated by treating the gel with medium containing nutrients and a steroid inducer. After such treatment, which could be repeated several times, the activity of the gels actually was higher than the initial activity (3–5 fold).

An interesting practical evaluation was carried out for a preparation of immobilized *C. simplex* with an approximate conversion capacity of 0.5 g of steroid per day per g of wet gel [186]. A continuously operated reactor loaded with only 1–2 kg of the immobilized *C. simplex* preparation would be sufficient to supply the Swedish demand of 1-dehydrogenated steroids (about 250 kg per year). This evaluation gives a general idea of the very modest reactor dimensions needed for the industrial production of such steroids.

In a similar study it was shown that polyacrylamide entrapped cells of various bacteria could transform Δ^5-3β-hydroxy-, Δ^5-3β-acetoxy-, and Δ^4-3-keto-steroids to $\Delta^{1,4}$-3-keto-steroids [187]. The products were obtained in 70–85% yield by employing immobilized cells of *Mycobacterium globiforme* or *C. simplex*. The immobilization of the bacterial cells reduced the rate of complete degradation of the steroids resulting in increased yield of the dehydrogenated compounds.

Whole cells of *C. simplex* have also been incorporated into collagen membranes for production of prednisolone from hydrocortisone [188]. Small chips of the collagen-cell complex were packed in a column and used for continuous operation. Maximum conversion was obtained at a substrate concentration of about 2mM and at a flow rate of SV = 2.

3β-Hydroxysteroid dehydrogenase was isolated from *Streptomyces griserocarneus* and immobilized on a copolymer of phenylenediamine and glutaraldehyde [189]. In the presence of Mg^{2+}-ions and substrate, 52% of the activity could be recovered after coupling. The preparation was used for production of Reichstein's compound S in a mixture of water and organic solvent, which was used to increase the solubility of the steroid compounds.

4.4 Organic Acids

The production of amino acids utilizing immobilized enzymes or microbial cells has been described above (Section 3.3). Amino acids have a wide applicability in the pharmaceutical industry and as pointed out earlier only the biologically active L-isomers are of interest. Of those discussed, citrulline, L-DOPA, and 5-hydroxytryptophan are of particular interest as medical agents. Other amino acids are important raw materials for the synthesis of various compounds as will be exemplified in this section.

4.4.1 α-Keto Acids

Recent research has indicated the potential use of α-keto acids for therapy of chronic uremia. A more thorough investigation of this therapeutic approach has been complicated by the difficulty in preparing the α-keto acid analogues of essential amino acids. Chemical synthesis is often complicated and elaborate, resulting in low yields. However, many α-keto acids can be prepared from the corresponding amino acid by an enzymatic reaction involving amino acid oxidase. Actually, for some keto acids this is the only realistic method of production. The cost of commercially available oxidases is high and therefore they need to be stabilized and re-used for as long a time as possible.

An immobilized two-enzyme system has been suggested for the production of α-keto acids [190]. The system consisted of an L-amino acid oxidase from rattlesnake (*Crotalus adamanteus*) venom and catalase immobilized on alkylamine glass. The oxidase was chosen because the keto-analogues of tryptophan and histidine, which cannot be prepared by any chemical synthesis method, are formed by the action of this enzyme.

The presence of catalase resulted in some advantageous effects. The oxidase was more stable, as well as more active, due to the degradation of denaturating hydrogen peroxide with the simultaneous replacement of utilizable oxygen. Furthermore, the risk of a secondary reaction between the keto acid and the accumulated hydrogen peroxide was reduced.

The studies were carried out using a packed bed reactor, but a tank reactor would probably be more effective from a kinetic point of view. The enzyme was inhibited by the substrate and, as pointed out above (Section 2.4.3), in such a case a tank reactor would be preferred. In addition to this, it would be easier to supply oxygen which seemed to be rate-limiting in the conversion of the amino acids to α-keto acids. The enzyme preparation was sufficiently active and stable for use in large scale and long-term experiments. Significant amounts of α-keto acids for use in clinical evaluations of the treatment of chronic uremia could thus be produced.

4.4.2 Urocanic Acid

Urocanic acid is used as a sun-screening agent and is produced from L-histidine by the action of microbial L-histidine ammonia-lyase. Of several microorganisms tested *Achromobacter liquidum* yielded the highest activity after immobilization by entrapment in polyacrylamide gel [39]. Before immobilization the cells were heat-treated at 70 °C for 30 min to inactivate urocanase, which converts urocanic acid to imidazolone propionic acid. This treatment did not affect the L-histidine ammonia-lyase activity of the cells. In contrast to free cells, the immobilized cells did not require a surfactant for maximum activity. Apparently some diffusion barrier for substrate and/or product was damaged upon immobilization.

The immobilized cells were used in a column for the continuous production of urocanic acid from histidine. An aqueous solution of histidine (pH 9.0) containing Mg^{2+}-ions (important for operational stability) was passed through the column at a flow rate of SV = 0.06 at 37 °C. Urocanic acid could easily be isolated from the effluent in more than 90% yield. The enzyme activity was very stable in the presence of Mg^{2+} (1 mM) and its half-life was 180 days at 37 °C [141]. It was concluded that this process was more advantageous for the commercial production of urocanic acid than the batch process using extracted enzyme or microbial broth suspensions.

All systems involving immobilized microbial cells described so far have been based on the entrapment of the cells in a gel lattice, i.e. polyacrylamide. Cells of *Micrococcus luteus* containing L-histidine ammonia-lyase activity have been immobilized by covalent coupling to carboxymethyl cellulose with carbodiimide [191]. The cells were fixed on the surface of the polymer, were they were in intimate contact with the substrate solution. The efficiency of the immobilized cells was therefore not reduced by any internal diffusion barrier. At least 75% of the enzyme activity remained after immobilization. Furthermore, it appeared that no damage to the cell membranes was encountered upon immobilization since the same dependence on surfactants was observed for free and immobilized cells. A permanent increase in membrane permeability was, however, obtained by an initial treatment of the immobilized cells with a surfactant. Urocanic acid could be continuously synthesized by passing a solution of histidine through a column packed with the immobilized cells. No decrease in activity was observed after 16 days operation at 23 °C.

4.4.3 L-Malic Acid

Tanabe Seiyaku Co. has developed a process for the continuous production of L-malic acid from fumarate [35]. The process involves cells of *Brevibacterium ammoniagenes* immobilized by entrapment in polyacrylamide. Approximately 15 t of the acid can be produced per month using a 1000 l column fed at a flow rate of 200 l per h of 1 M sodium fumarate. The column has been operating since 1974, although not continuously due to the small market for the product. Malic acid is used for treatment of hepatic malfunctioning, especially for hyper-ammonemia. It is also used as one of the components of amino acid infusions.

B. ammoniagenes was chosen from many microorganisms as having the highest fumarase activity. The utilization of the immobilized microbe was complicated by the production of succinic acid. Since it was difficult to remove this by-product from malic acid, attempts were made to suppress this side reaction. By treating the immobilized cells with bile extract the formation of succinic acid was reduced from 5 to less than 0.2 mole-% of malic acid. Compared to intact cells, the immobilized cells showed 75% of the activity after the bile extract treatment, though they showed only 5% of the activity immediately after immobilization. Obviously the bile extract treatment, in addition to suppressing the succinic acid formation, destroyed the membrane barrier for substrate and product transport. The stability of fumarase was increased considerably upon immobilization and the half-life was calculated to be 52 days at 37 °C. Furthermore, it was determined that the equilibrium of the fumarase reaction (= maximum conversion to malic acid, approximately 80%) was reached at 37 °C and a flow rate of SV < 0.23. Partially purified fumarase from *Pseudomonas* sp. was entrapped in cellulose triacetate fibers and used for the production of L-malic acid from fumarate [192]. A relatively high stability, probably due to local high protein concentration in the fiber, was observed, i.e. after continuous operation for 100 days about 70% of the initial activity remained. When the column was used as a recycling reactor at high flow rates an increased activity of the fibers was obtained. Obviously external diffusion of substrate was rate-limiting in the continuous flow process due to low linear velocity (see Section 2.3.2).

4.4.4 2-Keto-L-Gulonic Acid

2-Keto-L-gulonic acid is an important intermediate in vitamin C manufacturing. It can be produced by microbial oxidation from L-sorbose with L-sorbosone as an intermediate. In the first step L-sorbose dehydrogenase transforms L-sorbose to L-sorbosone, which is converted to 2-keto-L-gulonic acid by L-sorbosone oxidase in the second step. Cells of *Gluconobacter melanogenus* entrapped in polyacrylamide were effective in converting L-sorbose to L-sorbosone [193]. Since the sorbosone oxidase activity of these cells was relatively low, sorbosone was accumulated and, when a critical concentration was reached, the sorbose dehydrogenase was inactivated. By co-immobilizing cells of *Pseudomonas syringae,* containing high levels of the oxidase, and cells of *G. melanogenus,* an effective system for the conversion of sorbose to keto-gulonic acid was obtained [194]. However, problems with differences in optimum immobilization and reaction conditions as well as stability for the cell preparation have to be solved in order to produce a satisfactory catalyst. It appears that immobilization and operation separately will increase enzyme activities and stabilities, thereby making the production of keto-gulonic acid more efficient.

4.5 Fine Chemicals

Biocatalytic conversions are the most convenient methods available for the production of many complex biologically active compounds. These conversions often involve synthetic reactions requiring more than one enzyme as well as coenzyme. Whole microbial cells, which can supply the coenzyme regenerating systems, are employed rather than purified enzymes. The utilization of immobilized biocatalysts for this purpose has to some extent been explored. As this technology appears to be very suitable it will be of increasing importance.

4.5.1 Coenzyme A

The synthesis of coenzyme A (CoA) from pantothenic acid requires a multienzyme system consisting of five enzymes. Cells of *Brevibacterium ammoniagenes*, containing the multienzyme system, were immobilized by entrapment in cellophane tubing or polyacrylamide gel [36]. Other methods of immobilization resulted in deactivation of the cells. Furthermore, while entrapped intact cells required a surfactant for activity, entrapped dried cells were active without such an addition. Obviously, upon drying of the cells a membrane barrier for substrate transport (most likely for ATP) was destroyed. The preparations were tested in repeated batch as well as continuous packed bed operations. The polyacrylamide entrapped cells, for instance, had a half-life of about 5 days when used in a column at 37 °C. CoA was synthesized in a repeated batch procedure from sodium pantothenate, cysteine and ATP. The isolated product had a purity of 91%.

Cells of *Sarcina lutea* entrapped in polyacrylamide were likewise utilized for the production of CoA in a batch operation [195]. After five reactions for 20 h the CoA production was unchanged.

In both these studies relatively low yields of product were obtained but in spite of this it is interesting to note that a specific synthesis can be carried out by an immobilized intact multienzyme system.

4.5.2 Glutathione

The production of glutathione from glutamate, cysteine, glycine, and ATP with immobilized glutathione synthetase has been described in two patents [196, 197]. In both cases the enzyme was purified from *Saccharomyces cerevisiae* and immobilized covalently to a solid support. The yield of glutathione could be increased considerably (from 25 to 58%) by including immobilized carbamylphosphokinase (from *Streptococcus faecalis*) as well as cyanate and phosphate in the reaction mixture [196]. Furthermore the immobilized enzyme yielded much more glutathione than the free enzyme, i.e. 82 and 7% respectively [197].

4.5.3 Porphobilinogen

The pyrrole, porphobilinogen, can be enzymatically synthesized from two molecules of δ-aminolevulinic acid (ALA) by the action of δ-aminolevulinic acid dehydratase (ALA-dehydratase) as outlined in Fig. 7. Porphobilinogen is a precursor of porphyrins, chlorophyll, and the corrin ring of vitamin B_{12} and it is therefore an important compound for biochemical, clinical, and synthetic work. Methods based on immobilized enzyme for the preparation of substantial amounts of the pyrrole have been developed [198, 199].

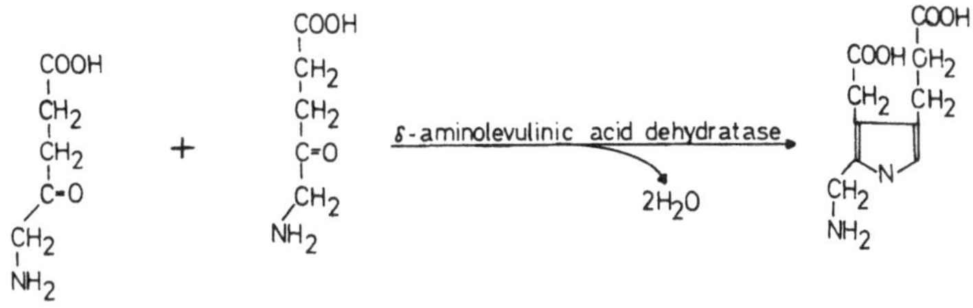

Fig. 7. Enzymatic synthesis of porphobilinogen from δ-aminolevulinic acid

ALA-dehydratase from bovine liver has been immobilized on BrCN-activated Sepharose [198, 200]. The enzyme is strongly product inhibited and therefore a column operation is most convenient. Thus, a column packed with the enzyme carrier complex was operated continuously for 9 days at 31 °C and at a flow rate of SV = 2.8 [198]. During this time 3.5 g of porphobilinogen was formed, which represented a yield of about 80%. Little, if any, loss of enzyme activity was observed during the experiments. Porphobilinogen has also been prepared in large scale quantities from ALA by the action of ALA-dehydratase fixed within the cells of *Chromatium vinosum* [199]. The cells were pretreated with ethanol to inactivate urogen synthetase in order to increase the yield of the product. In a typical preparation the cell mass (1.6 kg) was suspended in buffer (60 l) at 30 °C. After de-aerization and addition of an antifoam agent, the reaction was started by addition of ALA (30 g). The temperature was lowered to 15 °C after 2 h and the cells were removed by centrifugation. Porphobilinogen (12.5 kg) of 97% purity could be isolated from the supernatant fluid. The used cell mass was recycled and the lost activity (9-12%) was replaced by addition of fresh cell mass.

4.6 Radioactive Compounds

Pharmaceutical compounds labelled with short-lived isotopes are very useful since they can be detected by external (i.e. noninvasive) methods. While chemical synthesis is relatively slow and yields racemic mixtures, enzymatic synthesis is more rapid and specific. The product from enzymatic synthesis may, however, contain antigenic or pyrogenic substances. This complication can be overcome by immobilization of the enzymatic catalyst resulting in a product free of external protein.
^{13}N-L-Alanine (^{13}N half-life 10 min) was synthesized in 4 min with an immobilized two-enzyme system [201]. When soluble enzymes were used the same operation required 10 min. ^{13}N-L-Glutamic acid was synthesized in the first step by passing a solution of ^{13}N-ammonia and α-ketoglutarate through a column with immobilized glutamate dehydrogenase at a high flow rate. The enzyme was covalently bound to a N-hydroxysuccinimide derivative of silica beads. After addition of pyruvate to the eluate a second column containing immobilized transaminase was utilized to prepare ^{13}N-L-alanine. Higher yields were obtained using the immobilized enzymes (70–75%) as compared to the soluble system (50–60%). The enzymes were immobilized on separate supports in order to enable the isolation and utilization of ^{13}N-L-glutamic acid. Thus, labelled glutamic acid and alanine of pharmaceutical quality could be conveniently synthesized by employing immobilized enzymes.

5 Applications in Waste Treatment

The production of glucose from waste cellulose (Section 3.1.1) and hydrolysis of lactose
in whey (Section 3.2.1) are two examples where immobilized enzymes have potential
use for the elimination of disposal problems, as well as for the recovery of valuable
nutrients. Many other systems, based on immobilized biocatalysts, have been suggested
and investigated for similar purposes. These systems have an added importance due to
growing concern with the effects of waste on the environment.

It has been shown, for instance, that immobilized α-amylase can be used for the clarification of colloidal starch-clay suspensions, such as "white water" from paper mills
[202]. The solids in such waters are difficult to remove even with the aid of flocculants,
unless the starch is partially hydrolyzed with α-amylase. After amylase treatment the
solids readily settle out with the aid of alum. Since very large volumes of water have
to be treated, it would be advantageous to utilize an immobilized enzyme which could
be used over long periods of time. When soluble enzyme is used, it has to be destroyed
before the clarified water can be recycled and utilized in the paper manufacturing. An
additional advantage of an immobilized enzyme is that this step is eliminated.

α-Amylase (*Bacillus subtilis*) was immobilized in various ways, the most stable preparation being made by adsorption of the enzyme on Duolite S-30, which is a phenol-
formaldehyde resin with protein adsorbing properties [202]. After 40 days operation
no measurable loss of activity was observed. Alternatively, other preparations of acceptable stability could be made by covalent coupling of the enzyme to nylon or
acrylamide grafted cotton cloth. All these preparations of immobilized α-amylase
could degrade starch in wastewater streams and thereby significantly improve the
clarification procedure. A relatively short contact time between substrate and the
immobilized enzyme was required for this improved clarification.

The wastewaters from coal conversion processes contain relatively large amounts of
dissolved hydrogen sulfide and ammonia, as well as phenolic compounds and thiocya-
nates. The gases can be recovered by stripping procedures and the phenol by solvent
extraction. The "dephenolated" material may, however, still contain more than 50 ppm
phenol, which is well above the acceptable concentration for wastewater. The residual
hazardous phenol can be removed by activated sludge system utilizing a large tank or
pond in which the reactive microorganisms are maintained in dilute suspension by agitators or gas sparging. However, the concentration driving force for the detoxification
reaction is very low since in this single-stage batch reactor the concentration of pullu-
tant in the effluent is the same as in the reactor (i.e. very low). Extremely large reactor
volumes are therefore required.

An interesting bioreactor design has been investigated in order to provide a more
effecient detoxification system [203]. A strain of *Pseudomonas* was immobilized by
adsorption to anthracite coal by recycling a suspension of the live microorganism
through the reactor for a few hours. The attached cells multiplied rapidly and after
about one week a steady-state condition was reached. Additional growth resulted in
aqueous suspension of cells indicating that the carrier was saturated. Since a fluidized
bed reactor system was used this additional biomass production did not complicate
the operation of the process. In this case a tapered reactor was utilized which has some

interesting features such as an improved flow distribution especially at the inlet, as well as a wider range of operating conditions [204]. With a reactor residence time of only a few minutes the concentration of phenol was reduced to 10–50 ppb. The conversion rate in the fluidized bed appeared to be limited by available oxygen rather than by limitations in the biological system. When the oxygen content was increased by saturation of the feed stream (an oxygen pressure of 2.7 atm was used) about 10 g of phenol per day and liter bed volume could be converted. Typical conversion rates in a batch reactor are much less than 1 g per day. With the new fluidized bed system the reactor volume could thus be decreased considerably.

Utilization of similar tapered fluidized bed reactors has been suggested for other processes such as lactose hydrolysis and microbiological denitrification [204].

In the pesticide industry there is a need for the development of methods for the detoxification of by-products and wastewaters. Pesticide-hydrolyzing enzymes, which have been isolated from various microorganisms, can serve as catalysts for the degradation of pesticides. For instance, a crude extract of a mixed bacterial culture grown on the insecticide parathion, contained strong hydrolase activity not only for parathion but also for eight other organophosphate pesticides [205]. These nine compounds represent approximately 50% of the total amount of produced organophosphate pesticides. The enzyme is easily produced and extracted from the mixed culture.

Preliminary studies on the immobilization of the hydrolase activity to glass, supported the assumption that immobilized pesticide-hydrolyzing enzymes can be used on an industrial scale for detoxification since good operational stability under laboratory conditions was observed [206]. Based on these results from the laboratory, an industrial scale operation was evaluated. For treatment of 10 m^3 of wastewater per h, containing parathion (10 mg per ml), in a stirred 1 m^3 enzyme reactor, approximately 20 kg of catalyst would be required. To prepare this amount of catalyst (145 μg of protein per gram of support) 3 g of crude protein, derived from 6–10 l of cell broth, was needed. The flow rate was 166 l per min and the residence time 4.1 min. These calculations clearly demonstrate the potential use of immobilized pesticide-hydrolyzing enzymes on an industrial scale.

6 Concluding Remarks

Immobilized biocatalysts are at present utilized in various industrial processes. Most of these are on a relatively small scale and to date only one large scale operation, i.e. the isomerization of glucose to fructose, is in production. There are two potential candidates for a second large scale process, i.e. immobilized glucoamylase and β-galactosidase and it is likely that both of these processes will be commercially utilized in the near future. Indeed, immobilized glucoamylase will probably be introduced soon in the sugar industry in order to make the production of high fructose syrups from starch a totally continuous process. A continuous saccharification step would improve the overall process considerably since the liquification and isomerization processes are already operating continuously.

In this context, it might be worthwhile looking into the reason for success of the glucose isomerase process. High fructose syrup is a relatively new product and the immobilized enzyme process was actually developed concomitantly with this product. Thus there was no resistance from established technology to introducing the process as occured in the case of glucoamylase. In addition the favorable economics of the process have, of course, contributed to the success.

In conclusion, it appears more likely that immobilized enzyme systems will find, in many cases, application when the process and the product are developed simultaneously. Often, there is resistance to replace a well established process with a new one unless it shows certain clearcut advantages.

The interest in immobilized microorganisms has been increasing ever since it was shown that such preparations could be efficiently used for the production of specific substances and the fact that re-activation is possible by exposing the immobilized cells to a nutritional medium emphasized this interest even more [185]. It appears that immobilized microbial cells will find wider application as industrial catalysts particularly for the synthesis of complex biological compounds, which requires a multienzyme system and/or a cofactor regenerating system.

It has also been indicated in this review that there are several processes based on immobilized biocatalysts which have potential use on an industrial scale. In the future, many of these and other systems not yet investigated, can be envisaged as potentially useful for the manufacture of various products. As of now, most processes will probably be utilized in the food and pharmaceutical industries but may later find wider application in the chemical industry.

Acknowledgements

This work was supported by grants from the American Cancer Society (BC-60), the National Institute of Health (USPHS CA 11683), and the Swedish Natural Science Research Council (3896-001).

7 References

1. Nelson, J. H.: J. Dairy Sci. **58**, 1739 (1975)
2. Nelboeck, M., Jaworek, D.: Chimia **29**, 109 (1975)
3. Pitcher, Jr., W. H.: Catal. Rev. - Sci. Eng. **12**, 37 (1975)
4. Zaborsky, O. R.: Immobilized Enzymes. Cleveland, Ohio: CRC Press 1973
5. Weetall, H. H. (ed.): Immobilized Enzymes, Antigens, Antibodies, and Peptides. New York: Marcel Dekker 1975
6. Messing, R. A. (ed.): Immobilized Enzymes for Industrial Reactors. New York: Academic Press 1975
7. Mosbach, K. (ed.): Methods in Enzymology, Vol 44. New York: Academic Press 1976
8. Wingard, Jr., L. B., Katchalski-Katzir, E., Goldstein, L. (eds.): Applied Biochemistry and Bioengineering, Vol. 1. New York: Academic Press 1976
9. Hjertén, S.: J. Chromatogr. **87**, 325 (1973)
10. Hofstee, B. H. J.: Anal. Biochem. **52**, 430 (1973)
11. Dahlgren Cardwell, K., Axén, R., Bergwall, M., Porath, J.: Biotechnol. Bioeng. **18**, 1573 (1976)
12. Mosbach, K., Gestrelius, S.: FEBS Lett. **42**, 200 (1974)

13. Höjeberg, B., Brodelius, P., Rydström, J., Mosbach, K.: Eur. J. Biochem. 66, 467 (1976)
14. Axén, R., Ernback, S.: Eur. J. Biochem. 18, 351 (1971)
15. Nilsson, H., Mosbach, R., Mosbach, K.: Biochim. Biophys. Acta 268, 253 (1972)
16. Johansson, A.-C., Mosbach, K.: Biochim. Biophys. Acta 370, 339 (1974)
17. Dinelli, D., Marconi, W., Morisi, F.: In: Methods in Enzymology, Vol. 44. Mosbach, K. (ed.), pp. 227–243. New York: Academic Press 1976
18. Chang, T. M. S.: In: Methods in Enzymology, Vol. 44. Mosbach, K. (ed.), pp. 201–218. New York: Academic Press 1976
19. Gregoriadis, G.: In: Methods in Enzymology, Vol. 44. Mosbach, K. (ed.), pp. 218–227. New York: Academic Press 1976
20. van Leemputten, E., Horisberger, M.: Biotechnol. Bioeng. 16, 385 (1974)
21. Liu, C. C., Lahoda, E. J., Galasco, R. T., Wingard, Jr., L. B.: Biotechnol. Bioeng. 17, 1695 (1975)
22. Chang, T. M. S.: Biochem. Biophys. Res. Commun. 44, 1531 (1971)
23. Vieth, W. R., Venkatasubramanian, K.: In: Methods in Enzymology, Vol. 44. Mosbach, K. (ed.), pp. 243–263. New York: Academic Press 1976
24. Solomon, B., Levin, Y.: Biotechnol. Bioeng. 16, 1393 (1974)
25. Stanley, W. L., Watters, G. G., Chan, B., Mercer, J. M.: Biotechnol. Bioeng. 17, 315 (1975)
26. Richards, F. M., Knowles, J. R.: J. Mol Biol. 37, 231 (1968)
27. Lowey, S., Slayter, H. S., Weeds, A. G., Baker, H.: J. Mol. Biol. 42, 1 (1969)
28. Wolodko, W. T., Kay, C. M.: Can. J. Biochem. 53, 175 (1975)
29. Boundy, J. A., Smiley, K. L., Swanson, C. L., Hofreiter, B. T.: Carbohydr. Res. 48, 239 (1976)
30. Mosbach, K., Mattiasson, B.: Acta Chem. Scand. 24, 2093 (1970)
31. Mattiasson, B., Mosbach, K.: Biochim. Biophys. Acta 235, 253 (1971)
32. Srere, P. A., Mattiasson, B., Mosbach, K.: Proc. Natl. Acad. Sci. USA 70, 2534 (1973)
33. Martensson, K.: Biotechnol. Bioeng. 16, 567 (1974)
34. Chibata, I., Tosa, T., Sato, T.: Appl. Microbiol. 27, 878 (1974)
35. Yamamoto, K., Tosa, T., Yamashita, K., Chibata, I.: Eur. J. Appl Microbiol. 3, 169 (1976)
36. Shimizu, S., Morioka, H., Tani, Y., Ogata, K.: J. Ferment. Technol. 53, 77 (1975)
37. Slowinski, W., Charm, S. E.: Biotechnol. Bioeng. 15, 973 (1973)
38. Sato, T., Tosa, T., Chibata, I.: Eur. J. Appl. Microbiol. 2, 153 (1976)
39. Yamamoto, K., Sato, T., Tosa, T., Chibata, I.: Biotechnol. Bioeng. 16, 1601 (1974)
40. Laurence, R. L., Kittrell, J. R., Hultin, H. O.: Enzyme Technol. Dig. 2, 7 (1973)
41. Hultin, H. O., Laurence, R. L., Kittrell, J. R.: Enzyme Technol. Dig. 2, 156 (1973)
42. Horvath, C., Sardi, A., Solomon, B. A.: Physiol. Chem. Phys. 4, 125 (1972)
43. Linko, Y., Saarinen, P., Linko, M.: Biotechnol. Bioeng. 17, 153 (1975)
44. Weetall, H. H., Havewala, N. B., Garfinkel, H. M., Buehl, W. M., Baum, G.: Biotechnol. Bioeng. 16, 169 (1974)
45. Strumeyer, D. H., Constantinides, A., Freudenberger, J.: J. Food Sci. 39, 498 (1974)
46. Baum, G.: Biotechnol. Bioeng. 17, 253 (1975)
47. Solomon, B., Levin, Y.: Biotechnol. Bioeng. 17, 1323 (1975)
48. Corno, C., Galli, G., Morisi, F., Bettonte, M., Stopponi, A.: Die Stärke 24, 420 (1972)
49. Krasnobajew, V., Böniger, R.: Chimia 29, 123 (1975)
50. Lee, D. D., Lee, Y. Y., Reilly, P. J., Collins, Jr., E. V., Tsao, G. T.: Biotechnol. Bioeng. 18, 253 (1976)
51. Swanson, S. J., Emery, A., Lim, H. C.: J. Solid-Phase Biochem. 1, 119 (1976)
52. Martensson, K., Mosbach, K.: Biotechnol. Bioeng. 14, 715 (1972)
53. Martensson, K.: Biotechnol. Bioeng. 16, 579 (1974)
54. Takasaki, Y., Takahara, Y.: Japan. Kokai 76:70875 (1976)
55. Mandels, M., Sternberg, D.: J. Ferment. Technol. 54, 267 (1976)
56. Karube, I., Tanaka, S., Shirai, T., Suzuki, S.: Biotechnol. Bioeng. 19, 1183 (1977)
57. Suzuki, S., Karube, I.: Japan. Kokai 77:47986 (1977)
58. Thompson, K. N., Johnson, R. A., Lloyd, N. E.: U.S. Patent 3788945 (1974)

59. Zittan, L., Poulsen, P. B., Hemmingsen, S. H.: Die Stärke 27, 236 (1975)
60. Poulsen, P. B., Zittan, L.: In: Methods in Enzymology, Vol. 44. Mosbach, K. (ed.), pp. 809–821. New York: Academic Press 1976
61. Oestergaard, J., Knudsen, S. L.: Die Stärke 28, 350 (1976)
62. Nielsen, M. H., Zittan, L., Hemmingsen, S. H.: In: Chem. Eng. Changing World, Proc. Plenary Sess. World Congr. Chem. Eng. lst. Koetsier, W. T. (ed.), pp. 183–198. Amsterdam: Elsevier 1976
63. Hupkes, J. V., van Tilburg, R.: Die Stärke 28, 356 (1976)
64. Messing, R. A., Filbert, A. M.: J. Agric. Food Chem. 23, 920 (1975)
65. R. J. Reynolds Tobacco Co.: U. S. Patent 3645848 (1972)
66. Baxter Laboratories Inc.: Brit. Patent 1274158 (1972)
67. Park, Y. K., Toma, M.: J. Food Sci. 40, 1112 (1975)
68. Strandberg, G. W., Smiley, K. L.: Biotechnol. Bioeng. 14, 509 (1972)
69. Lee, Y.Y., Fratzke, A. R., Wun, K., Tsao, G. T.: Biotechnol. Bioeng. 18, 389 (1976)
70. Yokote, Y., Kimura, K., Samejima, H.: In: Immobilized Enzyme Technology. Weetall, H. H., Suzuki, S. (eds.), pp. 53–67. New York: Plenum Press 1975
71. Giovenco, S., Morisi, F., Pansolli, P.: FEBS Lett. 36, 57 (1973)
72. Korus, R. A., Olson, A. C.: J. Food Sci. 42, 258 (1977)
73. Kolarik, M. J., Chen, B. J., Emery, Jr., A. H., Lim, H. C.: In: Immobilized Enzymes in Food and Microbial Processes. Olson, A. C., Cooney, C. L. (eds.), pp. 71–83. New York: Plenum Press 1974
74. Linko, Y.-Y., Pohjola, L., Linko, P.: Process Biochem. 12(6), 14 (1977)
75. Taguchi, H., Suga, K., Yoshida, T., Yuda, S.: In: Immobilized Enzyme Technology. Weetall, H. H., Suzuki, S. (eds.), pp. 151–167. New York: Plenum Press 1975
76. Vieth, W. R., Venkatasubramanian, K.: In: Methods in Enzymology, Vol. 44. Mosbach, K. (ed.), pp. 768–776. New York: Academic Press 1976
77. Marconi, W., Gulinelli, S., Morisi, F.: Biotechnol. Bioeng. 16, 501 (1974)
78. Dickensheets, P. A., Chen, L. F., Tsao, G. T.: Biotechnol. Bioeng. 19, 365 (1977)
79. Kreen, M., Köstner, A., Kask, K.: Tr. Tallin. Politekh. Inst. 331, 131 (1973)
80. Suzuki, H., Ozawa, Y., Oota, H., Yoshida, H.: Agric. Biol. Chem. 33, 506 (1969)
81. Obara, J., Hashimoto, S.: Sugar Technol. Rev. 4, 209 (1976/77)
82. Saimaru, H., Izumi, C., Narita, S., Yamada, M.: Ger. Offen. 2518280 (1975)
83. Sato, K., Terashima, M.: Japan. Kokai 74:66886 (1974)
84. Reynolds, J. H.: In: Immobilized Enzymes in Food and Microbial Processes. Olson, A. C., Cooney, C. L. (eds.), pp. 63–70. New York: Plenum Press 1974
85. Messing, R. A.: In: Immobilized Enzymes in Food and Microbial Processes. Olson, A. C., Cooney, C. L. (eds.), pp. 149–156. New York: Plenum Press 1974
86. Markey, P. E., Greenfield, P. F., Kittrell, J. R.: Biotechnol. Bioeng. 17, 285 (1975)
87. Greenfield, P. F., Laurence, R. L.: J. Food Sci. 40, 906 (1975)
88. Bouin, J. C., Atallah, M. T., Hultin, H. O.: In: Methods in Enzymology, Vol. 44. Mosbach, K. (ed.), pp 478–488. New York: Academic Press 1976
89. Hasselberger, F. X., Allen, B., Paruchuri, E. K., Charles, M., Coughlin, R. W.: Biochem. Biophys. Res. Commun. 57, 1054 (1974)
90. Pappel, K. E., Köstner, A. I., Letunova, E. V., Tikhomirova, A. S.: Appl. Biochem. Microbiol. 12, 328 (1976)
91. Pitcher, Jr., W. H., Ford, J. R., Weetall, H. H.: In: Methods in Enzymology, Vol. 44. Mosbach, K. (ed.), pp. 792–809. New York: Academic Press 1976
92. Jakubowski, J., Giacin, J. R., Kleyn, D. H., Gilbert, S. G., Leeder, J. G.: J. Food Sci. 40, 467 (1975)
93. Portetelle, D., Thonart, Ph.: Lebensm. – Wiss. u. – Technol. 8, 278 (1975)
94. Kilara, A., Shahani, K. M., Wagner, F. W.: Lebensm. – Wiss. u. – Technol. 10, 84 (1977)
95. Kobayashi, T., Ohmiya, K., Shimizu, S.: In: Immobilized Enzyme Technology. Weetall, H. H., Suzuki, S. (eds.), pp. 169–183. New York: Plenum Press 1975

96. Pappel, K. E., Siimer, E. K., Köstner, A. I., Letunova, E. V., Tikhomirova, A. S.: Appl. Biochem. Microbiol. **12**, 173 (1976)
97. Morisi, F., Pastore, M., Viglia, A.: J. Dairy Sci. **56**, 1123 (1973)
98. Ohmiya, K., Ohashi, H., Kobayashi, T., Shimizu, S.: Appl. Environ. Microbiol. **33**, 137 (1977)
99. Pastore, M., Morisi, F., Zaccardelli, D.: In: Insolubilized Enzymes. Salmona, M., Saronio, C., Garattini, S. (eds.), pp. 211–216. New York: Raven Press 1974
100. Pastore, M., Morisi, F., Leali, L.: Milchwissenschaft **31**, 362 (1976)
101. Anonymous: Chem. Week **121**(20), 37 (1977)
102. Coughlin, R. W.: 4th Enzyme Engineering Conference, Bad Neuenahr, Germany 1977
103. Cheryan, M., van Wyk, P. J., Olson, N. F., Richardson, T.: J. Dairy Sci. **58**, 477 (1975)
104. Green, M. L., Crutchfield, G.: Biochem. J. **115**, 183 (1969)
105. Ferrier, L. K., Richardson, T., Olson, N. F., Hicks, C. L.: J. Dairy Sci. **55**, 726 (1972)
106. Cheryan, M., van Wyk, P. J., Olson, N. F., Richardson, T.: Biotechnol. Bioeng. **17**, 585 (1975)
107. Taylor, M. J., Cheryan, M., Richardson, T., Olson, N. F.: Biotechnol. Bioeng. **19**, 683 (1977)
108. O'Neill, S. P.: Biotechnol. Bioeng. **14**, 201 (1972)
109. Richardson, T., Olson, N. F.: In; Immobilized Enzymes in Food and Microbial Processes. Olson, A. C., Cooney, C. L. (eds), pp 19–40. New York: Plenum Press 1974
110. Chu, H. D., Leeder, J. G., Gilbert, S. G.: J. Food Sci. **40**, 641 (1975)
111. Wang, S. S., Gallili, G. E., Gilbert, S. G., Leeder, J. G.: J. Food Sci. **39**, 338 (1974)
112. Björck, L., Rosén, C.-G.: Biotechnol. Bioeng. **18**, 1463 (1976)
113. Lee, E. C., Senyk, G. F., Shipe, W. F.: J. Food Sci. **39**, 927 (1974)
114. Lee, E. C., Senyk, G. F., Shipe, W. F.: J. Dairy Sci. **58**, 473 (1975)
115. Swaisgood, H. E., Janolino, V. G., Horton, H. R.: In Advances in Enzyme Engineering, Vol. 2. Tsao, G. T. (ed.), pp 171–193. West Lafayette, Indiana: Purdue University 1976
116. Swaisgood, H. E.: Personal communication
117. Altomare, R. E.: U.S. NTIS, PB Rep., PB-245096 (1974)
118. Thananunkul, D., Tanaka, M., Chichester, C. O., Lee, T.-C.: J. Food Sci. **41**, 173 (1976)
119. Smiley, K. L., Hensley, D. E., Gasdorf, H. J.: Appl. Environ. Microbiol. **31**, 615 (1976)
120. Chibata, I., Tosa, T., Sato, T., Mori, T., Matuo, Y.: In: Fermentation Technology Today. Terui, G. (ed.), pp. 383–389. Osaka: Soc. Ferment. Technol. 1972
121. Chibata, I., Tosa, T., Sato, T., Mori, T.: In: Methods in Enzymology, Vol. 44, Mosbach, K. (ed.), pp. 746–759. New York: Academic Press 1976
122. Chibata, I., Tosa, T.: In: Applied Biochemistry and Bioengineering, Vol. 1. Wingard, Jr., L. B., Katchalski-Katzir, E., Goldstein, L. (eds.), pp. 329–357. New York: Academic Press 1976
123. Yokote, Y., Fujita, M., Shimura, G., Noguchi, S., Kimura, K., Samejima, H.: J. Solid-Phase Biochem. **1**, 1 (1976)
124. Bartoli, F.: 4th Enzyme Engineering Conference, Bad Neuenahr, Germany 1977
125. Tosa, T., Sato, T., Mori, T., Matuo, Y., Chibata, I.: Biotechnol. Bioeng. **15**, 69 (1973)
126. Tosa, T., Sato, T., Mori, T., Chibata, I.: Appl. Microbiol. **27**, 886 (1974)
127. Tosa, T., Sato, T., Nishida, Y., Chibata, I.: Biochim. Biophys. Acta **483**, 193 (1977)
128. Chibata, I.: 4th Enzyme Engineering Conference, Bad Neuenahr, Germany 1977
129. Matsui, M., Yoneya, T., Nagura, M.: Japan. Kokai 75:100285 (1975)
130. Fukui, S., Ikeda, S., Fujimura, M., Yamada, H., Kumagai, H.: Eur. J. Appl. Microbiol. **1**, 25 (1975)
131. Zaffaroni, P., Vitobello, V., Cecere, F., Giacomozzi, E., Morisi, F.: Agr. Biol. Chem. **38**, 1335 (1974)
132. Marconi, W., Bartoli, F., Cecere, F., Morisi, F.: Agr. Biol. Chem. **38**, 1343 (1974)
133. Chibata, I., Kakimoto, T., Nabe, K.: Japan. Kokai 74:81590 (1974)
134. Chibata, I., Kakimoto, T., Nabe, K.: Japan. Kokai 74:81591 (1974)
135. Fukumura, T.: Japan. Patent 74:15795 (1974)
136. Kanemitsu, O.: Japan. Kokai 75:132179 (1975)

137. Kanemitsu, O.: Japan. Kokai 75:132180 (1975)
138. Chibata, I., Tosa, T., Sato, T., Yamamoto, K.: Japan. Kokai 74:75782 (1974)
139. Chibata, I., Tosa, T., Sato, T., Yamamoto, K.: Japan. Kokai 75:100289 (1975)
140. Yamamoto, K., Sato, T., Tosa, T., Chibata, I.: Biotechnol. Bioeng. 16, 1589 (1974)
141. Chibata, I., Tosa, T., Sato, T.: In Methods in Enzymology, Vol. 44. Mosbach, K. (ed.), pp. 739–746. New York: Academic Press 1976
142. Royer, G. P., Andrews, J. P.: J. Macromol. Sci. Chem. A7, 1167 (1973)
143. Weetall, H. H.: In: Immobilized Enzymes for Industrial Reactors. Messing, R. A. (ed.), pp. 201–226. New York: Academic Press 1975
144. Venkatasubramanian, K., Saini, R., Vieth, W. R.: J. Food Sci. 40, 109 (1975)
145. Basarova, G., Turkova, J.: Brauwiss. 30, 204 (1977)
146. Wildi, B. S., Weeks, L. E.: U.S. Publ. Pat. Appl. B 430213 (1976)
147. Witt, Jr., P., Sair, R. A., Richardson, T., Olson, N. F.: Brew. Digest 45, 70 (1970)
148. Drawert, F., Hagen, W., Nitsche, T., Sipos, S.: Brauwiss. 27, 297 (1974)
149. Hebert, J. P., Scriban, R.: Brauwiss. 29, 365 (1976)
150. Lieberman, E. R.: Enzyme Technol. Dig. 4, 69 (1975)
151. Woodward, J. D., Bennett, A. B.: Brit. Patent 1421955 (1976)
152. Nunokawa, Y., Saito, J.: J. Soc. Brew. Japan 71, 286 (1976)
153. Gaina, B. S., Pavlenko, N. M., Datunashvili, E. N., Krylova, Y. I., Kozlov, L. V., Antonov, V. K.: Appl. Biochem. Microbiol. 12, 167 (1976)
154. Gaina, B. S.: Konserun. Ovoshchesush. Prom-st 1977, 10
155. Kuninaka, A., Sakaguchi, K.: Japan. Patent S. 36-5287 (1961)
156. Noguchi, S., Shimura, G., Kimura, K., Samejima, H.: J. Solid-Phase Biochem. 1, 105 (1976)
157. Ekström, B., Lagerlöf, E., Nathorst-Westfelt, L., Sjöberg, B.: Sven. Farm. Tidskr. 78, 531 (1974)
158. Lagerlöf, E., Nathorst-Westfelt, L., Ekström, B., Sjöberg, B.: In: Methods in Enzymology, Vol. 44. Mosbach, K. (ed.), pp. 759–768. New York: Academic Press 1976
159. Marconi, W., Cecere, F., Morisi, F., Della Penna, G., Rappuoli, B.: J. Antibiot. 26, 228 (1973)
160. Giacobbe, F.: 4th Enzyme Engineering Conference, Bad Neuenahr, Germany 1977
161. Savidge, T., Powell, L. W., Warren, K. B.: Ger. Offen. 2336829 (1974)
162. Cawthorne, M. A.: Ger. Offen. 2356630 (1974)
163. Hueper, F.: Ger. Offen. 2157972 (1973)
164. Hueper, F.: Ger. Offen. 2157970 (1973)
165. Brandl, E., Knauseder, F.: Ger. Offen. 2503584 (1975)
166. Heuser, L. J., Chiang, C., Anderson, C. F.: U.S. Patent 3446705 (1969)
167. Ryu, D. Y., Bruno, C. F., Lee, B. K., Venkatasubramanian, K.: In: Fermentation Technology Today. Terui, G. (ed.), pp. 307–314. Osaka: Soc. Ferment. Technol. 1972
168. Berezin, I. V.: Personal communication
169. Mandel, M. O., Köstner, A. I., Siimer, E. Kh., Kleiner, G. I., Elizarovskaya, L. M., Shtamer, V. Ya.: Appl. Biochem. Microbiol. 11, 197 (1975)
170. Kleiner, G., Elizarovskaya, L., Mandel, M., Köstner, A.: Tr. Tallin Politekh. Inst. 383, 31 (1975)
171. Marconi, W., Bartoli, F., Cecere, F., Galli, G., Morisi, F.: Agr. Biol. Chem. 39, 277 (1975)
172. Fujii, T., Hanamitsu, K., Izumi, R., Yamaguchi, T., Watanabe, T.: Japan. Kokai 73:99393 (1973)
173. Kamogashira, T., Kawaguchi, T., Miyazaki, W., Doi, T.: Japan. Kokai 72:28190 (1972)
174. Huber, F. M., Chauvette, R. R., Jackson, B. G.: In: Cephalosporin and Penicillin Compounds: Their Chemistry and Biology. Flynn, E. H. (ed.), pp. 27–73. New York: Academic Press 1973
175. Fujii, T., Matsumoto, K., Watanabe, T.: Process Biochem. 11 (8), 21 (1976)
176. Yamaguchi, T., Ishii, H.: Ger. Offen. 2331295 (1974)
177. Toyo Jozo Company: Belg. Patent 803832 (1973)
178. Fujii, T., Matsuda, T.: Japan. Kokai 75:157591 (1975)

179. Fujii, T., Shibuya, Y., Matsuda, T.: Japan. Kokai 75:116686 (1975)
180. Fujii, T., Shibuya, Y., Matsuda, T.: Japan. Kokai 75:116687 (1975)
181. Takeda, H., Matsumoto, I., Matsuda, K.: Japan. Kokai 75:3588 (1975)
182. Hueper, F.: Ger. Offen. 2409569 (1975)
183. Abbott, B. J., Cerimele, B., Fukuda, D. S.: Biotechnol. Bioeng. 18, 1033 (1976)
184. Konecny, J.: 4th Enzyme Engineering Conference, Bad Neuenahr, Germany 1977
185. Mosbach, K., Larsson, P.-O.: Biotechnol. Bioeng. 12, 19 (1970)
186. Larsson, P.-O., Ohlson, S., Mosbach, K.: Nature 263, 796 (1976)
187. Voishvillo, N. E., Kamernitskii, A. V., Khaikova, A. Ya., Leontev, I. G., Paukov, V. N., Nakhapetyan, L. A.: Izv. Akad. Nauk SSSR, Ser. Chem. 1976, 1303
188. Venkatasubramanian, K., Vieth, W. R., Constantinides, A.: 75th Annu. Meet. Am. Soc. for Microbiol., Session 116, 1975
189. Szentirmai, A.: 4th Enzyme Engineering Conference, Bad Neuenahr, Germany 1977
190. Fink, D. J., Falb, R. D., Bean, M. K.: In: Advances in Enzyme Engineering, Vol. 2. Tsao, G. T. (ed.), pp. 79–113. West Lafayette, Indiana: Purdue University 1976
191. Jack, T. R., Zajic, J. E.: Biotechnol. Bioeng. 19, 631 (1977)
192. Marconi, W., Morisi, F., Mosti, R.: Agr. Biol. Chem. 39, 1323 (1975)
193. Martin, C. K. A., Perlman, D.: Biotechnol. Bioeng. 18, 217 (1976)
194. Martin, C. K. A., Perlman, D.: Eur. J. Appl. Microbiol. 3, 91 (1976)
195. Chibata, I., Kakimoto, T., Nishimura, N., Nabe, K.: Japan. Kokai 75:126884 (1975)
196. Miwa, N.: Japan. Kokai 76:144789 (1976)
197. Miyamoto, I., Miwa, N.: Japan. Kokai 77:51089 (1977)
198. Gurne, D., Shemin, D.: In: Methods in Enzymology, Vol. 44. Mosbach, K. (ed.), pp. 844–849. New York: Academic Press 1976
199. Vogelmann, H., Ghahremani, B., Wagner, F.: Eur. J. Appl. Microbiol. 2, 19 (1975)
200. Stella, A. M., Wider de Xifra, E., Batlle, A. M. del C.: Mol. Cell. Biochem. 16 97 (1977)
201. Cohen, M. B., Spolter, L., Chang, C. C., Mac Donald, N. S., Takahashi, J., Bobinet, D. D.: J. Nucl. Med. 15, 1192 (1974)
202. Smiley, K. L., Boundy, J. A., Hofreiter, B. T., Rogovin, S. P.: In: Immobilized Enzymes in Food and Microbial Processes. Olson, A. C., Cooney, C. L. (eds.), pp. 133–147. New York: Plenum Press 1974
203. Scott, C. D., Hancher, C. W. Holladay, D. W., Dinsmore, G. B.: Second Symposium on Environmental Aspects of Fuel Conversion Technology, Hollywood, Florida, 1975
204. Scott, C. D., Hancher, C. W.: Biotechnol. Bioeng. 18, 1393 (1976)
205. Munnecke, D. M.: Appl. Environ. Microbiol. 32, 7 (1976)
206. Munnecke, D. M.: Appl. Environ. Microbiol. 33, 503 (1977)

Starch Hydrolysis by Immobilized Enzymes Industrial Applications

B. Solomon*
Department of Biophysics, The Weizmann Institute of Science
Rehovot, Israel

Contents

1 Introduction 132
2 Occurrence and Isolation of Starch 133
3 The Chemical Structure of Starch 134
   3.1 Amylose 134
   3.2 Amylopectin 135
   3.3 Minor Constituents of Starch 137
4 Enzymic Hydrolysis of Starch by Soluble Enzymes 138
   4.1 Hydrolysis of α-$(1 \rightarrow 4)$-D-Glucosidic Linkages 139
      4.1.1 α-Amylase $[(1 \rightarrow 4)$-α-D-Glucan Glucanohydrolase] (E.C. 3.2.1.1) 139
      4.1.2 β-Amylase $[\alpha$-D-$(1 \rightarrow 4)$ Glucan Maltohydrolase] (E.C. 3.2.1.2) 142
      4.1.3 Glucoamylase $[(\alpha$-1,4-Glucan Glucohydrolase)] (E.C. 3.2.1.3) 144
   4.2 Enzymic Degradation of $\alpha(1 \rightarrow 6)$-D-Glucosidic Linkages 145
      4.2.1 R-Enzymes $[(\alpha$-D-$(1 \rightarrow 4)$ $(1 \rightarrow 6)$ Glucan 6-Glucanonhydrolase)]
          (E.C. 3.2.19) 145
      4.2.2 Pullulanase 145
      4.2.3 Isoamylase $[\alpha$-D-$(1 \rightarrow 4)$ $(1 \rightarrow 6)$-Glucan 6-Gluconohydrolase] 146
      4.2.4 Amylo$(1 \rightarrow 6)$-Glucosidase (E.C. 3.2.1.33) 146
5 Industrial Applications–Use of Enzymes in the Conversion of Starch to Glucose 146
6 Hydrolysis of Starch by Immobilized Enzymes 150
   6.1 Immobilized Enzymes 150
   6.2 Enzyme Reactor Types, and the Appropriate Selection 151
   6.3 Preparation and Applications of Immobilized α-Amylase Derivatives 154
   6.4 Immobilized β-Amylase Preparations 160
   6.5 Preparation and Applications of Immobilized Amyloglucosidase Derivatives 163
   6.6 Immobilized Pullulanase 169
   6.7 Immobilized Two Enzyme System: β-Amylase and Pullulanase 170
7 Concluding Remarks 171
8 Acknowledgement 172
9 References 172

---

* Present address: Institute of Biochemistry and Biophysics, University of Tehran, Tehran, Iran

Starch, a mixture of two polysaccharides—amylase and amylopectin—is found mainly in the plant
kingdom. After cellulose, starch is the most widely commercially-utilized polysaccharide, being the
chief source of carbohydrate in the human diet. Starch and its degradation products have many uses
in industry, the food industry being the major consumer. By means of enzymic or acid hydrolysis a
range of syrups, having a wide variation in composition of glucose, maltose and higher oligosaccharides, can be produced. Enzymes generally used are α- and β-amylases and amyloglucosidase.
Recent development of the immobilization of enzymes, which permit a controlled process, easy
recovery and re-use of enzymes increases the potential of the enzymic hydrolysis of starch. Different preparations of immobilized starch-degrading enzymes and their performances in comparison
with soluble enzymes are discussed.

1 Introduction

The wide distribution of starch in plant-life, its great food value to both plants and animals and its extensive field of usefulness in domestic life and in many of the arts, sciences and trades, have made this substance a subject of study for generations.
The primary literature concerning starch is very voluminous [1–6]; many monographs
and extensive review articles dealing exclusively with starch have appeared since 1896
[7–9]. Within the last fifty years the chemical structure of starch has been elucidated
[10–13] and many of its physical and chemical properties have been correlated with the
high molecular weight nature of this important biopolymer [14–16].
Because of its wide distribution in nature, starch was used since an early period, not
only as food, but also as a useful product in various practical and industrial applications.
Strips of Egyptian papyrus, cemented together with a starch adhesive, have been dated
to the late Neolithic period [17]. From the fourteenth century onward, starch was a
common article of commerce. Because of the low cost of starch in comparison with
other natural products it can be put through certain well-designed chemical modifications to yield useful cheap products.
The range of starch chemical structure modifications and related use properties is
extremely wide. Over the years, a large number of relatively high molecular weight
products of starch with useful properties related to the polymeric nature of the polysaccharide have been prepared. The degradation products of starch represent the second
class of starch products and are mainly used as basic carbohydrate sweetners in the food
industry. They are obtained from starch that has been hydrolyzed to form water-soluble products ranging from maltodextrins to syrups and crystalline glucose known commercially as dextrose. Enzymic hydrolysis often provides advantages over nonenzymatic
hydrolysis. Because enzyme catalysis is quite specific it is often possible to avoid undesirable side reactions.
Recent developments in the production and use of immobilized enzymes which are
adsorbed or chemically bound to materials that are insoluble in the reaction medium,
further emphasize the inherent advantages of enzymic systems. The relatively high concentrations of enzyme in immobilized systems also permit the use of continuous reactors at relatively low temperatures with reasonably sized vessels. This results in the prolonged use of the enzyme which decreases the overall cost of enzyme per unit weight of
product. However enzyme engineering, especially the use of insolubilized enzymes in the
catalytic degradation of starch, has been rather slow getting off the ground.

2 Occurrence and Isolation of Starch

Starch, after cellulose, is perhaps the most widely distributed naturally occurring organic compound. Starch material has been found in some bacteria, protozoa and algae; its main sources, however, are in the higher plants where it may account for 20 to 70% of the weight. Starch occurs in various sites in the plant; in seeds such as cereal grains; in the root and tuber such as tapioca and potato; and more rarely, in the stem pith as in the sago plant. In the tubers roots, seeds and fruits of some plants, starch forms a semi-permanent reserve; but in leaf tissues the starch is transient and in content does not exceed 1–2%. Under the microscope, starch appears as minute rounded granules, whose size, shape and markings are specific for each species. In diameter the granules vary in the range of 15 to 40 μm (see Fig. 1) [18]. The starch granules represent a convenient insoluble source of energy, which can be gradually made available through the action of enzymes.

The starches can be subdivided into two groups according to their origin: starch originating from organs which retain a high water content and starch originating from organs with a relatively low water content. The first group includes starches that are deposited in tubers, bulbs, rhizomes, and stems. The second group includes the cereal starches [19]. The starch granules in most plant materials can be easily isolated [20] by extraction in a blender in the presence of mercuric chloride to inhibit enzymic activity; they readily sediment out from the extract. The resulting starch is contaminated by different fats and protein. Various purification procedures are described in the literature [21]. Despite the great variety of the starch supply, only a limited number of products are used for the manufacture of starch on an industrial scale. These include maize (corn

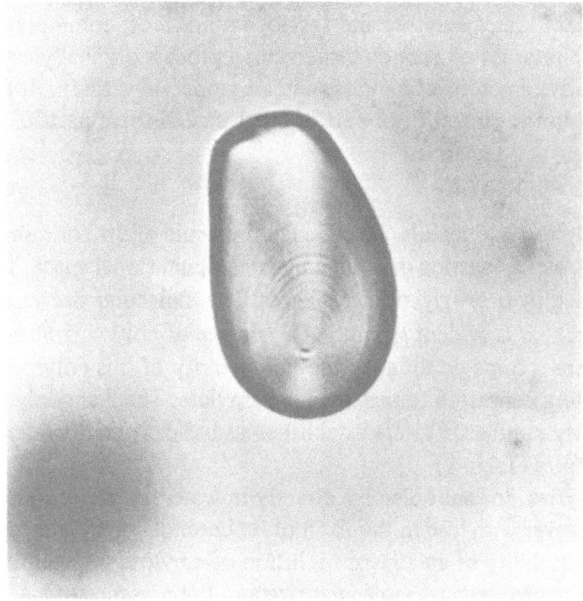

Fig. 1. A photograph of a potato starch granule ($\times$ 400) [18]

starch), potatoes (farina), manioc root (tapioca), wheat and rice. More recently, new types of starch have been developed on a large scale from waxy maize, sorghum and waxy sorghum.

The procedure used for the isolation of starch is governed by the source. The extraction and purification of starches from cereals is more difficult than is the case for many other botanical sources, presumably because the cereal grains usually contain considerably less moisture. Cereal starches require steeping—a preliminary swelling of the grain in water—whereas starches of noncereal sources require cell rupture or maceration of the tissue to free the starch. A general method for isolation of cereal starches was recommended by Adkins and Greenwood [22].

3 The Chemical Structure of Starch

The chemical structure of starch is essentially the same, regardless of the wide variety of sources from which it can be obtained. The essential features of the starch molecules as an α-D-$(1 \rightarrow 4)$-linked glucan with α-D$(1 \rightarrow 6)$-linked branched points were established by classical methylation and hydrolysis techniques [10–13, 23]. Before the 1940's starch was regarded as a single polysaccharide having a complex molecular architecture, but studies carried out in the U. S. A. by T. J. Schoch and in Switzerland by K. H. Meyer showed that starch was in fact a mixture of two polysaccharides. Both are polymers of α-D-glucopyranose; the major component-amylopectin—having a branched structure, whilst the minor component—amylose—is a linear macromolecule (see Fig. 2) [24]. In some plants, such as the potato, the amylopectin is esterified with a small amount of phosphate. The two components differ significantly in many physical properties, e.g. molecular size, solubility in water and iodine-staining power.

Separation of amylose and amylopectin is most commonly effected by selective partial precipitation of a starch dispersion, a process originally established by Schoch [25]. The relative amounts of amylose and amylopectin varies in different starches, but is generally in the range of one part of amylose and three parts of amylopectin.

3.1 Amylose

Amylose is a flexible, linear chain molecule of 500 or more glucose units which are capable of twisting or coiling in three dimensional space. The glucose residues are joined by α-1,4-glycosidic linkages. The molecular size varies with the source of the starch, as is evident from the DP (degree of polymerization) values given in Table 1. There is some doubt as to the full linearity of this polymer, since it cannot be completely converted to maltose by β-amylase. The β-amylolysis limit and the limiting viscosity number $[\eta]$ were established as basic criteria for the characterization of amylose samples (Table 1).

Amylose does not dissolve directly in water, or at best, dissolves only to a limited extent. However, amylose in the form of its butanol complex disperses into aqueous solution. The stability of an aqueous solution of amylose is dependent on several factors including pH, concentration, molecular size and the presence of electrolytes. The α-D-$(1 \rightarrow 4)$ bond

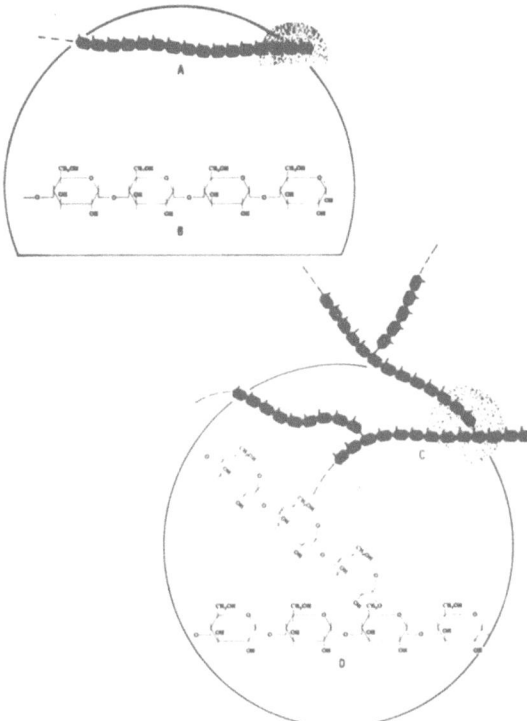

Fig. 2. Structure of the amylose and amylopectin components of starch. (A) diagram of a portion of an amylose molecule; (B) enlarged view of the shaded section showing chemical formula; (C) diagram of a portion of an amylopectin molecule; (D) enlarged view of shaded area showing chemical formula [24]

enables the molecule to attain various conformations in solution. The variation in molecular conformations in solution is responsible for "retrogradation" and "gelatinization" of concentrated amylose solutions. The term retrogradation has been employed in the starch field to describe the process whereby starch in the dissolved or hydrated state reverts to a water-insoluble form [26]. From measurements of the weights of amylose remaining in solutions during retrogradation, Whistler and Johnson [26] derived that the rate of retrogradation for several types of starches occurs in the order: potato < corn < wheat. The complex of amylose with iodine producing a blue color is the basis for the quantitative determination of amylose. Potentiometric iodine-titration measurements carried out with iodine, and the apparent iodine-binding capacity of amylose has been widely studied [27–30]. Generally the amperometric and potentiometric methods were used assuming that pure amylose binds 19–20% iodine on a weight basis. The iodine complex is associated with a helical structure of the starch chains, since dextrins containing less than six D-glucose residues do not produce the characteristic colored complex [31–33]. More details of the chemical and physicochemical analysis of the molecular structure of amylose are given elsewhere [34–37].

3.2 Amylopectin

The constitution of amylopectin is undoubtedly more complex than that of amylose, particularly, because of its highly branched structure. Amylopectin is also a glucose

Table 1. Properties of the amylose components (according to Greenwood and Thomson [20])

Starch	Iodine affinity	β-Amylolysis limits[a]			Approx.
		(i)	(ii)	[η]	D.P.
Grasses					
Amylomaize	19.2	77	101	180	1300
Barley	19.0	73	100	250	1850
Oat	19.2	77	–	180	1300
Wheat	19.1	68	–	280	2100
Zea mays	18.8	78	100	150	1100
Leguminosae					
Broad bean	19.2	82	99	240	1800
Pea (smooth-seeded)	19.2	81	100	180	1300
Pea (wrinkled-seeded)	19.2	82	101	140	1000
Underground storage organs					
Irish rhizome	19.1	84	100	240	1800
Parsnip	19.4	72	99	590	4400
Potato	19.5	76	100	410	3000
Miscellanous fruits and seeds					
Apple	19.0	84	99	200	1500
Banana	19.9	82	100	240	1300
Mango kernel	19.2	77	100	240	1800
Hevea (endosperm)	19.2	79	101	220	1600
Hevea (cotyledon)	19.0	74	100	200	1500

[a] Percentage conversion into maltose with (i) pure β-amylase and (ii) a mixture of β-amylase and Z-enzyme.

polymer, having a configuration composed of linear chains similar to those of amylose, but connected at the branch points by α-1,6-linkages. These branch points are believed to occur at intervals of about 20–30 glucose units and some evidence indicates that there may also be some branching at C(2) and C(3). The molecular weight of amylopectin is estimated to be in the millions. Most of the available molecular weight data on amylopectin from various sources have been summarized by Greenwood [37]. Recently Whelan *et al.* [38, 39] have described a new enzymic method of chain-length assay. The results obtained by these authors are in agreement with those obtained by the classical methods [40, 41]. Although the ramified nature of amylopectin is well established, the exact architecture of the molecule has not yet been defined. Various branched structures have been proposed [21, 41–45] as illustrated in Fig. 3. Relatively few physical studies on amylopectins, other than molecular weight measurements, have been reported. The branched structure and enormous molecular weight distribution seriously impede any attempts at interpretation of these data. Amylopectin forms reasonable solutions in water. Neutral aqueous solutions of amylopectin are extremely stable and there is little tendency for the molecules to "retrograde", although retrogradation may occur to a limited extent at low temperature. Regarding the reaction

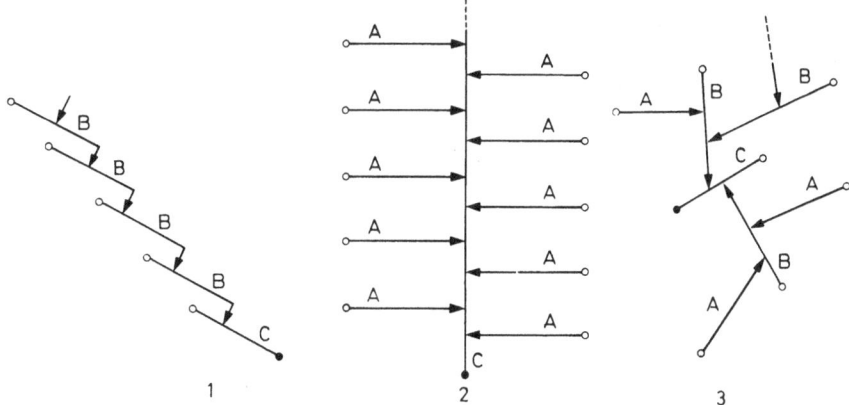

Fig. 3. Schematic representation of proposed structure for the amylopectin component. o = terminal, nonreducing end groups; • = reducing end group; → = α-D-(1 → 6) linkage; — = chain of 20 to
25 α-D-(1 → 4)-linked D-glucose residues. 1–Structure proposed by Haworth [40]. 2–Structure
proposed by Staudinger and Husemann [42]. 3–Structure proposed by Meyer and Bernfeld [43].
All of these structures contain different arrangements of the same basic units. A-chains are side
chains linked solely by their reducing group to the rest of the molecule; B-chains are those to which
A-chains are attached, although they themselves are similarly linked by their reducing group to
another chain; and C-chains carry the reducing groups [21, 45]

with iodine, amylopectin forms a red coloration with iodine in contrast to the blue color
obtained with amylose. Potentiometric titrations show that the proportion of iodine
bound by amylopectin in solution is considerably smaller than that bound by amylose
under the same conditions. It has been suggested [24] that this low binding power is due
to the fact that the branch points disturb the helix formation of the amylopectin molecule.

3.3 Minor Constituents of Starch

Starches commercially available are produced from different sources and usually contain
minor amounts of noncarbohydrate constituents [46]. These constituents, often called
"contaminants" in the industry, originate either from the raw material from which the
starch has been derived or as a result of the chemical treatment of the plant tissues to
increase manufacturing yields. Among the contaminants, other than water, one should
list higher fatty acids, other lipids and inorganic compounds of which phosphoric acid
has attracted considerable attention [47–48]. Lampitt and coworkers [49] confirmed
earlier findings that most of the phosphorus of potato starch is covalently bound to the
starch and that the amylopectin contains more esterified phosphate than amylose. In
wheat starch, however, phosphorus is present, either largely or wholly in the form of
adsorbed phosphatides preferentially associated with the amylose fraction. These findings were confirmed and extended by other authors [50, 51].
The first systematic analysis of the fatty substances in starch was made by Taylor and
Nelson [52]. They found that the major fraction of the fatty material could not be re-

moved by common fat solvents before acid hydrolysis. Lehrman [53] suggested that the fatty acids might be present as an adsorption complex with the carbohydrate moiety. This explanation was supported by adsorption experiments in which palmitic acid was added to various defatted starches.

4 Enzymic Hydrolysis of Starch by Soluble Enzymes

As outlined earlier, starch is the most abundant storage reserve carbohydrate in the plant kingdom and is also an important raw material for numerous commercial products. The degradation of starch by enzymes is of great biological and industrial importance. Starch hydrolases degrade the polysaccharide to soluble, low molecular weight products which undergo well known metabolic processes. Because of the information now available on the specificity of the various starch-splitting enzymes, they have become important tools in the elucidation of the structure of starch. The importance of the starch-degrading enzymes in industry stems from their application in the preparation of cereal beverages, in bread making and in the modification of starches for use as sizes, adhesives and in many other applications. Enzymes which promote the hydrolysis of starch to reducing sugars were detected over a century ago in a variety of biological materials. They are widely distributed in nature occurring in the digestive secretions of animals and within the cells of most animals, plants and microorganisms. The group name originally given to the enzymes catalysing starch hydrolysis was *diastase*. As information on the starch hydrolysing enzymes has accumulated this term was replaced by *amylases*. They act on starch and derived polysaccharides to hydrolyze the α-1,4-glycosidic linkages. The amylases may be divided into three groups: the α-amylases which split the bonds in the interior of the substrate (endoamylases); the β-amylases which hydrolyze units from the nonreducing end of the substrate (exoamylases) and the glucoamylases which split off glucose units from the nonreducing terminal of the substrate molecules. With the recent emphasis on the nomenclature of enzymes; α-D-(1 $\rightarrow$ 4)-glucan-4-glucanohydrolase, α-D-(1 $\rightarrow$ 4)-glucan maltohydrolase and α-1,4-glucan glucohydrolase have been proposed for the above types of amylases, respectively. The salient features of the action of these enzymes are illustrated in Fig. 4, representing the attack of the three types of enzyme on amylose and amylopectin [54].

Another group of starch hydrolases are the debranching enzymes which hydrolyse the $\alpha(1 \rightarrow 6)$-D-glucosidic linkages in amylopectin and α-dextrins and form a relatively newly discovered class of carbohydrases. This group of enzymes may be divided into: direct debranching enzymes—pullulanases and isoamylase—and indirect debranching enzymes—amylo-1,6-glucoamylases [55]. The direct debranching enzymes hydrolyze the α-1,6-bonds of unmodified starch amylopectin. The action of the "indirect" debranching enzymes on the other hand must be preceded by a modification of the substrate with another enzyme.

The elucidation of the action patterns of the various hydrolases on starch has been greatly facilitated by the availability of techniques such as paper chromatography and Sephadex column chromatography. These techniques used in conjunction with the older methods such as the increase in reducing power or the decrease in viscosity, or

Fig. 4. Schematic diagram showing amylose action. (A) Random degradation of amylose by α-amylase to form dextrins. (B) Gradual degradation of amylose by β-amylase from nonreducing end to form maltose. (C) Hydrolysis of amylose by glucoamylase to yield glucose. (D) Random degradation of amylopectin by α-amylase to form dextrins. (E) Gradual degradation of amylopectin by β-amylase from nonreducing ends to form maltose. (F) Degradation of amylopectin by glucoamylase to yield glucose [54]

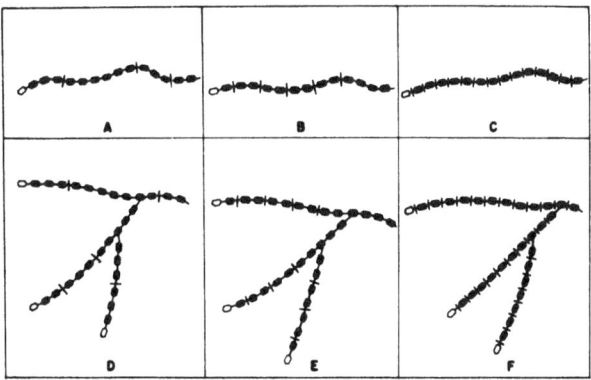

iodine-staining ability, give valuable information about the action patterns of the enzymes under discussion.

4.1 Hydrolysis of α-(1 → 4)-D-Glucosidic Linkages

4.1.1 α-Amylase [(1 → 4)-α-D-Glucan Glucanohydrolase] (E.C.3.2.1.1)

The α-amylases are calcium metalloproteins and are widely distributed in nature. Several α-amylases have been crystallized [56–58]. The molecular weights of α-amylases appear to be in the order of 50,000 with each molecule containing 1 gram atom of Ca^{2+} [59]. Calcium is bound to the enzyme molecule very tightly and is required for enzymic activity and for conformational stabilization. Removal of the calcium renders the enzymes susceptible to proteolysis [60]. It was shown that the strength of the binding of calcium ions to the protein varies according to the source of the enzyme, but for all α-amylases its presence increases the stability of the enzyme toward denaturation by heat, acid or urea [60]. Amylases give typical bell-shaped curves when activity is plotted against pH. The maximum activities of the amylases are in the acid region between pH 4.5 and 7.0, but the shape of the activity curves and location of the pH optima differ depending on the enzyme source. The α-amylases, which have a full complement of calcium are less heat-labile than the β-amylases. This property is of great importance in food processing, e.g. bread baking. It is also significant that α-amylases from different sources vary in their heat stability.

A. Action Patterns of the Various α-Amylases
The α-amylases, found virtually in every type of living cell, effect a rapid fragmentation of the whole starch molecule by cleaving the α-D-(1 → 4) linkage of the polymer more or less at random, and bringing about a slow conversion of starch to reducing sugars.
The action of α-amylase on the amylose fraction of starch proceeds in two stages. Ini-

tially, a complete, rapid degradation of amylose into maltose and maltotriose takes place. This step of α-amylolysis is essentially the result of a random attack on the substrate by the enzyme. Typical of this breakdown is a rapid loss of viscosity and of the iodine-staining power of the amylose. The second stage [61] is much slower than the first step, involving a slow hydrolysis of the oligosaccharides with the formation of glucose and maltose as the final products.

The α-amylolysis of amylopectin yields glucose, maltose and a series of α-limit dextrins, oligosaccharides of four or more glucose residues, all containing α-1,6-glycosidic bonds Further hydrolysis of the products resulting from the first stage of enzymolysis proceeds slowly, effecting a breakdown of certain linkages in the regions of the branch points of the molecule.

The liquefying power is determined by measurements of the changes brought about in a standard potato starch gel during a definite period of hydrolysis. Starch is rapidly altered by α-amylases in such a way that the solution is no longer colored by iodine. This stage in the degradation is designated as *the achroic point* and the reducing value calculated as maltose is termed the achroic R-value. The time required for starch to be hydrolyzed to the achroic point, or a definite color just prior to it, may be used as a practical measure of the enzymic activity. The α-amylases from different sources do not show identical action on starch. In the following paragraphs some of the action patterns of α-amylases of different origin will be described.

a) α-Amylases from Animal Sources. Salivary amylase and pancreatic amylase produced in the salivary gland and the pancreas are the principal α-amylases of animal origin. Like all other α-amylases, they promote a more or less random fragmentation of the starch molecule by hydrolyzing the α-D-($1 \rightarrow 4$) glucosidic bonds in the inner and outer chains of the compound. Considerable amounts of low molecular weight reducing sugars, identified as maltose, maltotriose and maltotetraose, are produced at the initial stages of hydrolysis [62–64]. The α-amylases cannot hydrolyze the α-D-($1 \rightarrow 6$) bonds of amylopectin. Starch fragments which are resistant to hydrolysis and contain the α-D-($1 \rightarrow 6$) linkage have been called *limit dextrins*. The α-amylase limit dextrins are relatively low molecular weight compounds; the exact chemical nature of which is not known [61]. Although the low molecular weight branched D-glucosyl oligosaccharides are not hydrolyzed by α-amylases, the linear compounds are readily hydrolyzed. By use of linear D-glucosyl oligosaccharides labeled with C^{14} at the reducing moiety [65], differences have been observed in the rate of hydrolysis of the D-glucosidic bond in the oligosaccharides studied (see Fig. 5). Nevertheless, one should keep in mind that the products of complete hydrolysis by the α-amylases from animal sources are maltose and D-glucose.

b) α-Amylases of Plant Origin. α-Amylases in plants are of importance for the conversion of starch into reducing sugars during the germination of seeds. The plant amylases are of particular value in the brewing, distilling and baking industries. The enzymes act in a similar manner in the case of animal α-amylases. The α-D-($1 \rightarrow 4$) linkages in the inner and outer chains of the starch are hydrolyzed at random, whereas the α-D-($1 \rightarrow 6$) and the α-D-($1 \rightarrow 3$) linkages cannot be hydrolyzed. Maltose cannot be hydrolyzed by plant α-amylases. Since D-glucose appears in large amounts during digestion and since maltotri-

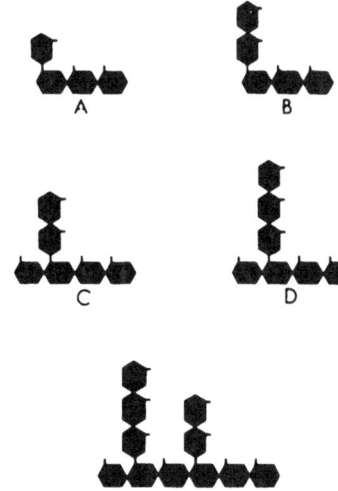

Fig. 5. Possible structures of oligosaccharides with α-D-(1 → 4) and α-D-(1 → 6) linkages (limit dextrins) produced from amylopectin by salivary amylase. (A) tetrasaccharide; (B) pentasaccharide; (C) hexasaccharide; (D) heptasaccharide; (E) undeconsaccharide [24]

ose is slowly hydrolyzed by the enzyme, plant α-amylases seem to hydrolyze terminal D-glucosidic bonds of starch [66].

c) α-Amylases of Microbial Origin. The α-amylases of a microbial source are the basis for many commercial applications in which hydrolytic enzymes are used. The enzymes have been obtained in crystalline form from bacterial sources *Bacillus stearothermophilus* [67], *Bacillus subtilis* [68, 69] and from fungal sources *Aspergillus niger* and *oryzae* [70, 71]. the microbial α-amylases effect a rapid fragmentation of starch by cleaving the α-D-(1 → 4) bonds in the inner and outer chains of the molecule. The enzymes cannot hydrolyze α-D-(1 → 6) or α-D-(1 → 3) bonds. Although the α-D-(1 → 4) linkage in maltose is not hydrolyzed, those in higher molecular weight maltooligosaccharides are hydrolyzed. With respect to the other α-amylases; D-glucose, maltose and a limit dextrin characteristic of each enzyme, are the final products of the digestion of starch by microbial α-amylases.

B. Mechanism of Enzymic Hydrolysis

It has been proposed [72] that the catalytic hydrolysis of starch occurs on the surface of the various α-amylases. Initially, some of the functional groups at the active site of the enzyme molecule interact with the functional groups of the substrate molecules to form the enzyme-substrate complex. The latter is then catalytically attacked by the carboxylate and imidazolium groups of the enzyme [72, 73]. Some possible arrangements of the functional groups of the substrate and enzyme in the complex are shown in Fig. 6. In the complex, the D-glucosidic bond oxygen has been protonated by hydrogen ions from amino or imidazole groups of the enzyme and the electron deficient center at C-1 of the bond attracts electrons from donor groups such as hydroxyl groups. The resulting structure is cleaved at the C-1 carbon side of the bond [72] forming a carbonium ion intermediate and a neutral D-glucosyl fragment. The final step involves the addition of a hydroxyl ion (or a water molecule) to the carbonium ion intermediate.

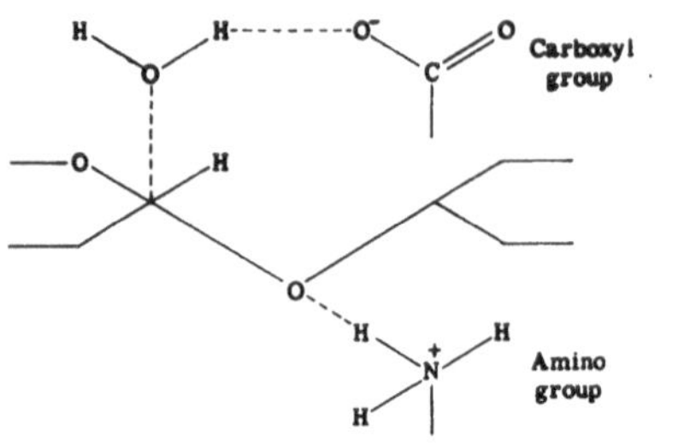

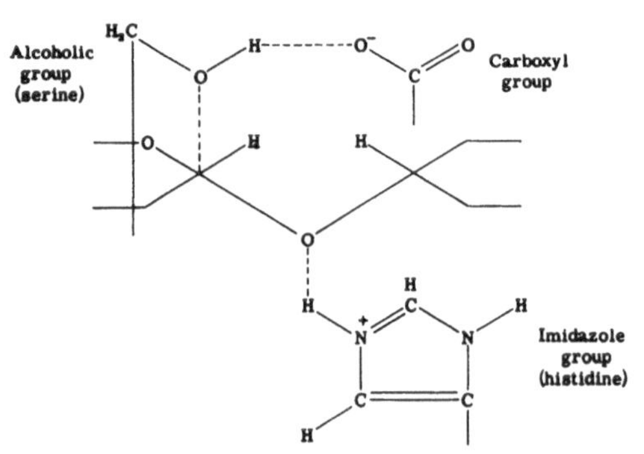

Fig. 6. Possible arrangements of functional groups at the active center of the enzyme and a D-glucosidic bond of starch [24]

4.1.2 β-Amylase [α-D-(1 → 4) Glucan Maltohydrolase] (E.C.3.2.1.2)

β-Amylase occurs in a great variety of plants. Plant sources include barley, wheat, rye, sweet potatoes and soybeans. Crystalline β-amylase was prepared initially from the sweet potato [74] and in recent years from many other plant sources [75]. The molecular weight of sweet potato β-amylase is approximately 150,000 [76]. The most active pH ranges of β-amylase are between pH 5.0 and 6.0 and they are completely stable between pH 4 and pH 8–9 at 20 °C for at least 24 h [77–80]. In contrast to the case of α-amylases, there is little evidence to suggest that the properties and action pattern of β-amylase are dependent on the enzyme source. The temperature of optimum activity is about 45 °C. Sulfhydryl groups are essential for enzymic activity and any reagent interacting with these causes inactivation [81]. The results of studies of the chemical structure of β-amylase have recently been reported [82].

A. Mechanism of Hydrolysis

β-Amylases hydrolyze the α-1,4-glycosidic bonds in starch by way of an inversion of the configuration with respect to the C(1) position of the glucose from α to β. This configurational change is the reason for the name given to the enzyme; the β-prefix does not signify that the enzyme recognized the β-glycosidic linkage.

It has been shown that one carboxyl, one imidazole and at least one sulfhydryl group are present at the active site of the enzyme and are involved in its catalytic activity [83].

Several investigators have indicated in schematic fashion, possible arrangements of the enzyme and substrate in the complex to account for the fact that maltose is the only hydrolytic product and that the enzyme cannot bypass an α-D-(1 → 6) linkage. The initial proposals of Myrbäck [84] have been extended by Thoma and Koshland [83] to the induced-fit concept. In this concept, the substrate induces a conformational change in the enzyme and the reaction groups are brought into close proximity for reaction with the substrate. If the nonterminal D-glucose residues of the starch chains are held at the active site, a conformational change is not possible and an unreactive complex is formed. A rearrangement of this complex must occur for the substrate to be hydrolyzed. With compounds such as the cyclic dextrins, their molecular size will not allow the interaction of the functional groups of the enzyme and the substrate, so that the cyclic dextrins cannot be hydrolyzed by β-amylase. Verification of the induced-fit concept for β-amylase awaits further experimentation.

B. Action Pattern of β-Amylases

β-Amylase starts at the nonreducing ends of the outer chains of starch and proceeds by gradual removal of maltose units.

Amylose with an even number of D-glucose units is converted completely to maltose, whereas amylose with an odd number of D-units is converted to maltose and to maltotriose. At high concentrations of enzyme and prolonged incubation, maltotriose is slowly hydrolyzed to D-glucose and maltose [24]. Amylopectin is hydrolyzed like amylose, beginning at the nonreducing end of the outer chains. Since β-amylase cannot hydrolyze an α-D-(1 → 6) bond, a high molecular weight limit dextrin containing all the original α-D-(1 → 6) linkages is thus produced [85]. Several possible structures for the outer chains of the β-amylase limit dextrin have been suggested by Summer and French [86] and are shown in Fig. 7. The attack of β-amylase on the substrate may be the result of: (a) a single chain mechanism in which the enzyme, having formed a complex with a substrate molecule, hydrolyzes that molecule completely, before attacking a second

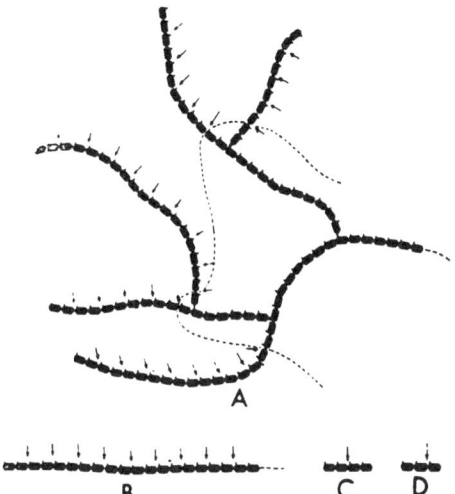

Fig. 7. Action pattern of β-amylase on amylopectin (A); amylose (B); maltotetraose (C) and maltotriose (D). Dotted arrow indicates slow rate of hydrolysis; solid arrows indicate rapid rate of hydrolysis; dotted line indicates the limit of the hydrolysis of amylopectin [24]

molecule of substrate–as was concluded by Cleveland and Kerr [87]; (b) a multichain mechanism in which the enzyme acts randomly on all the substrate molecules as suggested by Hopkins [88] and (c) a multiple attack mechanism in which the enzyme splits off several maltose molecules with the substrate molecule, and then diffuses away to combine with another molecule of substrate.

During single-chain action, only maltose and undegraded substrate molecules will be present in the system because the turnover number of β-amylase is extremely high [76, 89]. In contrast during multichain attack all the amylase molecules will be degraded by approximately the same amounts. Investigations of beta amylolysis of small substrate molecules (D. P. < 50) seem to suggest that the action takes place by multiple attack [90, 91]. Although conflicting evidence has been presented for the mechanism of attack by β-amylase on large substrate molecules, most workers favor the multichain mechanism [92].

4.1.3 Glucoamylase (α-1,4-Glucan Glucohydrolase) (E.C. 3.2.1.3)

Glucoamylase is an exo-splitting enzyme that removes the glucose units consecutively from the nonreducing ends of the starch polymers. Other names that have been used in the literature for this type of enzymic activity include amyloglucosidase, glucamylase and λ-amylase. Preparations of glucoamylase may be easily obtained from culture filtrates from several fungal species of the *Aspergillus* and *Rhizopus* groups [93–96] and from certain yeasts and bacteria [97, 98]. Crystalline glucoamylase from several fungal sources have also been described [99, 100]. The molecular weight of the glucoamylase from *Aspergillus niger* is 97,000 [101] and the pH optimum lies in the range of pH 4 to 5. The enzyme is relatively heat stable and has an optimum temperature of activity at 50° to 60 °C for up to 24 h.

Glucoamylase hydrolyzes amylopectin, amylose and the maltooligosaccharides completely to D-glucose [100]. Isomaltose [102] and nigerose [103] are also hydrolyzed demonstrating that the enzyme is capable of hydrolyzing the α-D-(1 → 6) and the α-D-(1 → 3) as well as the α-D-(1 → 4) bond. Rate measurements of the hydrolysis of maltose, nigerose and isomaltose have shown that the relative rates of hydrolysis of these compounds are in the ratio of 30 to 3 to 1 [103]. Radioactive oligosaccharides labeled at the reducing end have been used to show conclusively that the enzyme acts predominantly by a multichain mechanism [104]. That the enzyme action proceeds by removal of single D-glucose units from the nonreducing ends has been shown in a number of laboratories [94, 104–107]. The finding that the D-glucose liberated in the reaction is of β-configuration must be taken into account in any action mechanism to be proposed for this enzyme [107].

The hydrolysis of the amylopectin component by glucoamylase begins at the nonreducing ends of the molecule and proceeds with a progressive shortening of all the chains. The hydrolysis of amylose also begins at the nonreducing end and proceeds with a progressive shortening of all the amylose chains until they are completely hydrolyzed to D-glucose. The ability of the glucoamylase to hydrolyze the α-D-(1 → 4) linkage in maltose is in marked contrast to the action patterns of the α and β amylases.

4.2 Enzymic Degradation of α-(1 → 6)-D-Glucosidic Linkages

Debranching enzymes hydrolyze the α-(1 → 6)-D-glucosidic linkages in amylopectin and α-dextrins. The study of these enzymes is important for the understanding of *in vivo* degradation of starch. They also provide an important tool for the investigation of the fine structure of starch. Their existence in plant extracts as well as in protein extracts from animal, yeast and bacterial tissues has been described [108, 109].

4.2.1 R-Enzymes [α-D-(1 → 4) (1 → 6) Glucan 6-Glucanohydrolase] (E. C. 3.2.19)

The R-enzyme was one of the first enzymes described [110] as capable of hydrolyzing the α-D-(1 → 6) linkages of starch. A crystalline preparation of the enzyme has not been obtained and little information is available on its physical properties.
R-enzymes were discovered in beans and potatoes [110] and recently in sweet corn [111] and have shown to have the ability to hydrolyze the α-1,6 glycosidic bonds in amylopectin and its β-limit dextrin. They also cleave 1,6-bonded α-maltose and maltotriose residues in α-limit dextrin and pullulan [112], but are unable to remove 1,6-linked α-glucose units [113].

4.2.2 Pullulanase

Pullulanases are enzymes of microbial origin. The activity and specificity of these enzymes are similar to those of the R-enzymes. However, the enzymes from various sources differ in the rates of hydrolysis of the substrates and in the composition of the degradation products, as is evident in the data given in Table 2 [55].
Pullulan (poly-α-1,6-malbacterium) produced by *Pullularia pullulans* consists of α-(1 → 6) linked maltotriose units is an ideal substrate for estimation of the activity of pullulanase since it is not degraded by the α- or β-amylases. The degradation of this substrate is presently believed to proceed by an endo-pattern [114]. The pullulanases are active on amylopectin, β-dextrins and on other polysaccharides [114].

Table 2. Relative rates of hydrolysis of polysaccharides by *Aerobacter* and sweet corn pullulanase (according to Lee and Whelan [55])

Substrate	Relative initial rates of hydrolysis		
		Sweet corn	
	A. aerogenes	F1	F2
Pullulan	100	100	100
Amylopectin β-dextrin	54	213	164
Amylopectin	17	13	12
Glycogen β-dextrin (shellfish)	34	7	7
Glycogen (shellfish)	–	–	1.6
Glycogen α-limit dextrins	54	213	152

4.2.3 Isoamylase [α-D-(1 → 4) (1 → 6)-Glucan 6-Gluconohydrolase]

These enzymes were isolated from yeasts and fungi [115–117]. The main difference between pullulanases and isoamylases is the inability of the latter enzymes to degrade the linear polysaccharide pullulan. Isoamylases debranch amylopectin, limit dextrins and glycogen. There is also a strict requirement for the α-1,6-linkage to constitute a true branch point, rather than a linkage in a linear chain. This last property makes this enzyme a valuable tool in studies of the carbohydrate structure.

4.2.4 Amylo(1 → 6)-Glucosidase (E.C. 3.2.1.33)

The debranching effected in an indirect fashion is the result of a two component enzyme system, amylo-1,6-glucosidase/oligo-1,4 → 1,4-glucan transferase [118] as illustrated in Fig. 8.

5 Industrial Applications—Use of Enzymes in the Conversion of Starch to Glucose

Patent literature contains numerous references dealing with the optimal conditions of the enzymic-hydrolysis of starch, with the various processes and reactors, which might be used for starch conversion, as well as with the industrial utilization of the resulting products. Some of the widely used industrial amylases are listed in Table 3. [119]. The table also contains the primary action of the enzymes and the practical use of the final products. The table shows that the food industry is one of the principal consumers of the enzymic hydrolysates of starch. Sugars and sweet syrups used commercially are usually produced by the controlled enzymic hydrolysis of starch [119–135]. The products are classified according to the content of reducing-sugar, representing D-glucose, and defined as D. E. or dextrose equivalent. D-glucose can also be prepared from starch by acid hydrolysis; in such a hydrolysis, however, yields are low and reversion reactions

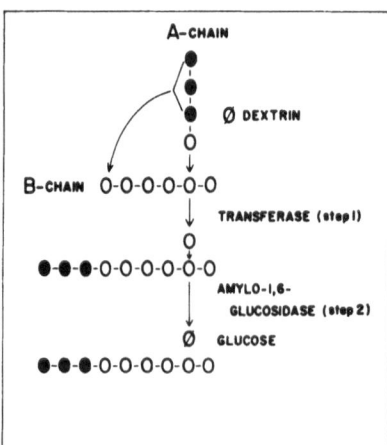

Fig. 8. Debranching of limit dextrin by glucosidase-transferase. The first step involves a glycosyl transfer, three glucose units being transferred from the side (A) chain to the main chain (B). This is followed by release of the single 1,6-linked α-glucosyl residue by the action of the amylo-1,6-glucosidase component of the enzyme [118]

seem to be considerable. In industry, starch is usually hydrolyzed by a mixture of amylolytic enzymes to yield a mixture of short chain oligoglucosides [123, 130, 133, 134]. The final product of hydrolysis, dextrose, is produced by a two-stage enzyme process [123]: in the first stage starch is liquified by α-amylase, the liquefying enzyme, whereas in the second stage the oligosaccharides formed are hydrolyzed to dextrose by amyloglucosidase—the saccharifying enzyme.

Table 3. Widely used industrial enzyme preparation marketed in the United States [119]

Trade name, source	Producer's code name	Principal and additional enzymes	Primary action on substrate	Utilization
Amyliq, bacterial	Wal	α-Amylase	Liquefies starch	Adhesives, sizings, paper coating
Amyzyme R, *Bacillus subtilis*		α-Amylase	Liquefies starch	Paper coating and wrap sizing, adhesives, convoyer belt cleaning
Animal diastase, hog pancreas glands	Cud[a]	α-Amylase, protease. lipase	Liquefies starch	Desizing agent in textile industry, sewage disposal
Clarase, fungal	M	α-Amylase, protease, phosphatase, maltase	Hydrolyzes starch	Fruit juices, chocolate syrup, brewing
Dextrinase, fungal	M	β-Amylase, protease amyloglucosidase	Converts dextrins to maltose and dextrose	Cereal syrups— higher DE
Diastase 50, pork pancreas	Wil	α-Amylase, amylopsin, animal diastase	Hydrolyzes starch	Digestive aid, pharmaceuticals
Diastase, 73 microbial	R & H	Amyloglucosidase, β-amylase	Saccharification of starch	Production of dextrose and dextrose syrups from starch
Diastatic malt syrup, malt	Std	Amylases	Hydrolyzes starch	Baking industry
Diazyme, micorbial	M	Amyloglucosidase	Hydrolyzes starch to glucose	Dextrose manufacture, sweetener fermentation, pharmaceuticals
Disposzyme, microbial	Wal	Protease, amylase, cellµlase	Mixture of enzymes obtained from selected microorganisms grown on bran and a special strain of bacteria	To aid in the clearing of household drain lines
Drain cleaner, formulation	M	Amylase, protease, lipase and anaerobic bacteria	Liquefies sewage	Cleaning drain lines, sewage treatment

Table 3 (continued)

Trade name, source	Producer's code name	Principal and additional enzymes	Primary action on substrate	Utilization
Enzygans, pancreas		Amylase, lipase	Hydrolyzes	Food products
Enzyme 4511-3, bacterial	Wal	Amylase, protease	Hydrolyzes starch and proteins	Crackers
Enzyme W-3-F, bacterial	Wal	α-Amylase (heat resistant)	Hydrolyzes starch	Starch, syrups with low sugar content
Exsize	PMP	α-Amylase, protease	Hydrolyzes starch	Desizing of textiles
Fermex, *Aspergillus oryzae*	Wal[a]	Amylase, protease	Hydrolyzes starch	Bakery, wafers
Fungal amylase, *Aspergillus oryzae*	M	α-Amylase, protease	Hydrolyzes starch to dextrins	Baking
Fungal protease, *Aspergillus oryzae*	M	Protease, α-amylase	Hydrolyzes proteins, starch	Baking, meat tenderizing
Gelatinase, formulation	M	Gelatinase, α-amylase	Solubilizes gelatin	Photography
HT-44 and HT-440, bacterial	M	α-Amylase, protease	Liquefies starch	Textiles, adhesives, paper, brewing, grain alcohol
HT-2000, bacterial	M	α-Amylase, β-amylase protease	Liquefies and saccharifies starch	Pharmaceuticals
Malt	Malting companies	α-Amylase, β-amylase	Converts starch to maltose	Brewery, distillery, bakery
Malt syrup diastic	Several malting companies	α-Amylase, β-amylase	Hydrolyzes starch	Baking
Milezyme A	M	α-Amylase	Hydrolyzes starch	Foodstuff for poultry, beef, cattle, swine
Mylase SA (Allase SA), fungal	Wal	α-Amylase, limit dextrinase, other carbohydrases	Hydrolyzes starch and its degradation products	Hydrolyzes converted starch syrups to dextrose syrups
Mylase SC (Allase SC), (a speciel form of mylase SA)	Wal	α-Amylase, limit dextrinase, other carbohydrases with added buffer salts	Hydrolyzes starch and its degradation products	Hydrolyzes converted starch syrups to dextrose syrups
Mylase 100, *Aspergillus oryzae*	Wal[a]	α-Amylase	Liquefies starch	Pharmaceuticals, digestive aid
Rhozyme S, *Aspergillus oryzae*	R & H[a]	α-Amylase, β-amylase, amyloglucosidase	Hydrolyzes starch	Syrup production
Rhozyme T-22, *Aspergillus oryzae*	R & H[a]	α-Amylase, β-amylase, amyloglucosidase	Hydrolyzes starch, dextrins	Syrup production from acid hydrolyzed starch
Rhozyme 33, *Aspergillus oryzae*	R & H	Amylase (high), low protease	Liquefies and saccharifies starch	Baking

Table 3 (continued)

Trade name, source	Producer's code name	Principal and additional enzymes	Primary action on substrate	Utilization
Vanzyme and Vanzyme L	V	α-Amylase	Liquefies starch	Food ingredients, adhesives, starch, syrups
Wilzyme 400, beef pancreas	Wil[a]	Amylase, protease, lipase	Hydrolyzes proteins, starch and fat, microorganisms	All-purpose digestive enzyme, pharmaceuticals, research
Zymo-best, formulation	PMP	Protease, amylase, gumase	Hydrolyzes proteins, starch	Feed ingredients for beef and dairy cattle, sheep, swine, poultry
Zypanar, pancreas	A	Protease, amylase, lipase	Hydrolyzes starch, proteins, fat	Digestive aid

[a] Products are available in several preparations of various enzymic strengths

Major producers of industrial enzymes in the United States	*Code used in Table*
Armour Pharmaceutical Co. (Reheis Chemical Co.), Chicago, Ill.	A
Cudahy Packing Co., Omaha, Neb.	Cud
Miles Chemical Co., Elkhart, Ind.	M
Premier Malt Products, Inc., Milwaukee, Wis.	PMP
Rohm & Haas Co., Philadelphia, Pa.	R & H
J. E. Siebel Sons Co., Chicago, Ill.	Sie
Standard Brands Scales Co., New York, N. Y.	Std
Wallerstein Co., Staten Island, N. Y.	Wal
Wilson Laboratories, Chicago, Ill.	Wil

In the standard industrial procedure, starch is enzymatically hydrolyzed by soluble amylases using a batch process. Several proposals are to be found in the literature, however, in which continuous processes for enzyme hydrolysis of starch on a large scale are described. Such a system for the continuous enzymatic degradation of starch has been recently proposed by Marshal and Whelan [136] (Fig. 9). In the system described, a starch solution is continuously added from a reservoir to an ultrafiltration cell by means of positive pressure from a compressed nitrogen tank. The enzyme, amyloglucosidase, is added to the ultrafiltration cell (C) and the product formed passes through a permeable membrane into a collecting vessel (E). The membrane is of sufficiently low porosity for the enzyme and undegraded starch to be retained in the cell [136]. Ultrafiltration techniques, however, seem to be of limited commercial value because of the low surface area to volume ratio in the reaction cell. In a commercial process this ratio should obviously be maximized.

Tubular membrane reactors, in which the enzyme is enclosed, fulfill this requirement. There is no wonder, therefore, that such reactors for starch hydrolysis have been de-

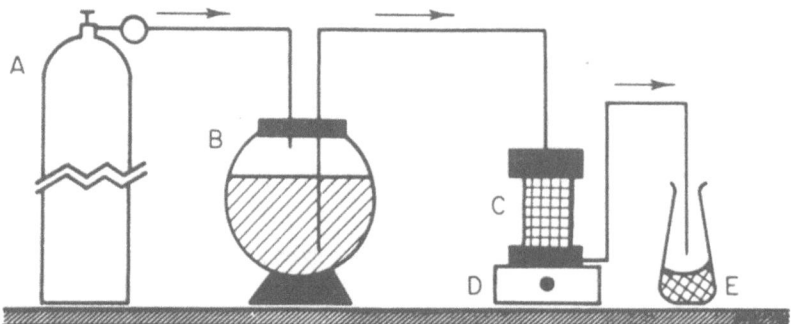

Fig. 9. Diagram for the continuous enzymatic conversion of starch to maltotetraose by *Pseudomonas stutzeri* amylase: (A) compressed nitrogen gas; (B) 3 l of 2% starch reservoir; (C) 400 ml reaction vessel, ultrafiltration cell with a PM-10Amicon membrane filter; (D) magnetic stirrer; (E) product collection vessel [136]

scribed in the literature [137–139]. Tachauer *et al.* [137] describe the hydrolysis of starch by a mixture of α- and β-amylases in a tubular membrane reactor proposed previously by Closset [138]. When used for prolonged time intervals, the system shows a better performance as a reactor in comparison with that of a solid wall reactor. A membrane with a sharp molecular weight cut-off is needed to improve the performance of proposed membrane reactor systems at short time intervals. The findings of Tachauer [137] and Closset [138] were recently confirmed by Subramanian [139].

The procedures developed for the binding of enzymes to insoluble carriers have opened up new possibilities for the continuous conversion of starch. Immobilized enzymes seem to be attractive for commercial use, especially in food processing as indicated by Olson and Richardson in their review article [140].

6 Hydrolysis of Starch by Immobilized Enzymes

6.1 Immobilized Enzymes

In the last decade a wide range of methods for the immobilization of enzymes and their subsequent characterization has been described in the literature [141–147]. The development of these techniques has produced novel forms of biochemical catalyst, which can be used in reactors to carry out specific chemical reactions of applied interest. At present, three enzymes at least—penicillin amidase, aminoacylase and glucose isomerase—are being used in industry in immobilized form on a large scale. Their successful commercial operation is encouraging, and there are many other reactions of interest to industry, where an immobilized enzyme could be used as the catalyst.

Among the most important enzyme immobilization techniques one should mention the following:

1. Covalent crosslinking of enzyme to enzyme without the benefit of carrier.
2. Crosslinking of enzyme within carriers or on the surface of carriers.
3. Covalent attachment to carriers.
4. Adsorption on or in carriers.
5. Encapsulation or entrapment.

Regardless of the manner of immobilization, water insoluble enzymes possess the following noteworthy characteristics: they can be separated from the reaction mixture containing the product and any residual reactants and they can be reused as active, specific catalysts in subsequent reactions. Significant and favorable considerations in the application of immobilized enzymes also include greater stability of the bound enzyme, enzymic activity over broad ranges of pH and temperature, development of continuous processes as well as the adaptability of immobilized enzymes to a variety of specific processes. Despite the very large number of methods of immobilization reported in the literature, many techniques are unsuitable for production of large quantities of immobilized enzyme, either because of the difficulty to scale-up or the very high cost of the carrier or the techniques employed. When choosing the most adequate immobilization procedure one must take into consideration not only the cost of the carrier but also the stability of the final immobilized enzyme. An expensive method of immobilization is likely to be justified for a cheap enzyme only if the operational life of the enzyme is very long or there are important process advantages.

The choice between use of the soluble free enzyme and the corresponding immobilized enzyme depends on the cost of the enzyme, the nature of the conversion process and the relative operational stabilities of the two forms. Some food processes, for example, involve the addition of the required enzymes at the final processing stage, making re-use impossible. In other instances, the ability to remove the immobilized enzyme from the product stream, ensuring minimal contamination by protein, may influence the choice. In all cases, it seems that the main factor determining the choice of the form of biocatalyst is the operational stability of the enzyme. If the enzyme can be stabilized by immobilization, then re-use of the enzyme is as a rule, worthwile. Not all of the immobilization techniques lead to enzyme stabilization; the study of the factors determining enzyme stability will thus undoubtedly be of continued interest to the bioengineer and biotechnologist.

6.2 Enzyme Reactor Types, and the Appropriate Selection

With the ease of enzyme recovery and re-use brought by immobilization, there has been considerable interest in studying the performance of immobilized enzymes in a wide variety of reactor types. These may be classified according to mode of operation and the flow pattern in the reactor. The most common system is the stirred tank normally operated as a batch system, but also semicontinuously by repeatedly drawing off part of the reaction liquor at intervals and refilling with fresh substrate solution. Some examples of continuous flow systems are shown in Fig. 10. The flow patterns range from the "well-mixed" continuous-flow stirred tank reactor (CSTR) to the ideal plug-flow or tubular reactor. The basic characteristics of these "ideal" reactor systems are summarized in Table 4 [148]. Several types of reactors where the flow pattern approximates that of a plug-flow reactor have been used. One variation is a reactor consisting of semipermeable

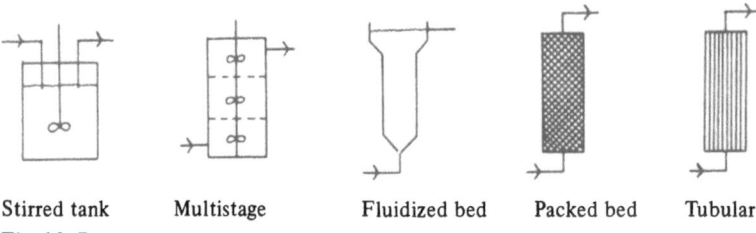

Stirred tank   Multistage   Fluidized bed   Packed bed   Tubular

Fig. 10. Reactor types

Table 4. The characteristics of "ideal" enzyme reactor systems
(according to Lilly and Dunnill [148])

Type of reactor	Flow pattern in reactor	Residence time distribution of liquid in reactor	Diagrammatic representation	Basic equations describing performance+
Batch	well-mixed	$N(t)$ vs t (reaction-time)	$S_o \rightarrow S$	$XS_o - K_m \ln(1-X) = \dfrac{kE \cdot t}{V}$
Tubular	Plug-flow	$N(t)$ vs $t/\bar{t}$	$q, S_o \rightarrow q, S$	$XS_o - K_m \ln(1-X) = kE \cdot 1/q$
CSTR	well-mixed	$N(t)$ vs $t/\bar{t}$	$q, S_o \quad q, S$	$SX_o + K_m \cdot \dfrac{X}{1-X} = kE \cdot 1/q$

* $N(t)$ is the number of elements of liquid in the reactor with a residence
time t, and $\bar{t}$ is the mean residence time of liquid leaving the reactor ($= V/q$).
+ Derived for an enzyme acting on a single substrate where the reaction rate
is described by the Michaelis-Menten equation, kE is the total enzyme activity
in the reactor (volume, V), and X in the proportion of the initial sustrate converted to product ($= S_o-S/S_o$).

hollow fibers inside which the substrate solution passes through the reactor with the enzyme solution in the space surrounding the fibers [149]. Alternatively, a tube may be
packed with a membrane sheet containing immobilized enzyme wound in a spiral about
spacers placed centrally down the tube [150]. In addition to the widely used packed beds
of immobilized enzyme particles or fibers, other types of porous bed reactors, which make
use of porous sheets or blocks, have been tried [151]. When the substrate solution is
pumped upward through a particle bed, it is possible to operate with the bed in an expanded state [152]. If the flow rate is increased, the system becomes fluidized and the
flow pattern changes to one intermediate between the two ideal flow patterns. When
using immobilized enzyme supports in the form of porous sheets or blocks, membrane,
tubers or fibers (that are an integral part of the reactor), no retention system is required.
Particles of immobilized enzymes may also be recovered from the product stream by
filtration or centrifugal sedimentation.
In the design of a reactor the amount of catalytic activity per unit volume of reactor is
an important factor, since this determines the size of reactor. The reaction rate per unit
volume will be a function of the weight of immobilized enzyme preparation per unit
volume of reactor, the protein content of the preparation, the specific activity of the

immobilized enzyme protein and the efficiency of the utilization of that activity. The nature of the reaction will sometimes impose particular requirements that influence the choice of reactor. For most biochemical reactions it is essential to control both the operating temperature and pH. This can be done without difficulty in a well mixed batch or continous reactor. The choice of the reactor system will also depend on other factors, such as utilization and cost. The required output from the reactor will influence the choice of mode of operation. Where output is small, continuous operation may be uneconomic. Batch operation, especially of stirred tank reactors, gives a relatively cheap and flexible system that can be used for a wide range of processes. Continuous operation usually requires the design of a reactor specifically for a particular process and will involve greater capital expenditure. However, continuous operation has several advantages: diminished labor costs; the ease of automatic control and the constancy of the reaction conditions. All the criteria outlined above require knowledge of the specific process. However, the effect of the reaction kinetics on the choice of the reactor could be described in general terms. It is usual to compare the three basic ideal reactor systems; the batch stirred tank-reactor, the continuous tubular reactor and the well-mixed continuous stirred tank reactor. The basic equations describing the performance of these reactors, assuming the enzyme obeys the Michaelis-Menten equation are given in Table 4. It is convenient, therefore, to compare directly the productivity of the tubular and CST reactors. This is illustrated in Fig. 11 [153] in which the ratio of the amounts of enzyme needed in the two reactors (E_{CST}/E_{PF}) is plotted against the conversion (X) for various S_0/K_m where S_0 is the feed substrate concentration. When $S_0 \gg K_m$ the two equations become identical, but when $S_0 \ll K_m$ the relative performance of the two reactors are

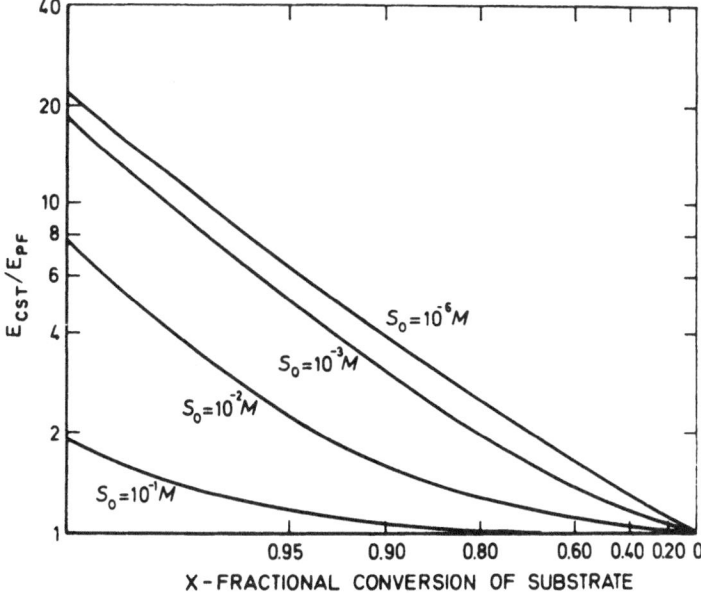

Fig. 11. Variation in the ratio (E_{CST}/E_{PF}) with conversion (X) for various values of feed substrate concentration (S_0) at the same K_m $(= 10^{-3}$ M) and constant flow rate [153]

different. The use of the Michaelis-Menten equation as a kinetic model simply may be
too inaccurate for a number of reactor designs or analysis studies. In an extensive study
of the use of glucoamylase immobilized on glass beads for converting starch to dextrose,
Weetall and Havewala [154] used both a packed bed and a stirred-tank reactor. They
examined the effects of flow-rate, pore size, immobilization techniques and concluded
that the simple Michaelis-Menten model was not sufficient to explain their experimental
system. Flow-dependent kinetic parameters of amyloglucosidase bound to cellulosic
support to convert maltose to glucose were observed [152] also with the packed-bed and
stirred tank operation. The packed-bed system appeared to follow first-order rather than
Michaelis-Menten kinetics. However, in the preceding comparison it was assumed that
the enzyme was not inhibited by either substrate or products of the reaction. Since many
enzymes are subject to inhibition of some kind, it is necessary to consider the effect of
inhibition on the kinetics of reactors. When the enzyme is subject to inhibition by excess
substrate, this is most serious in a batch or plug-flow system but the inhibition may be
reduced by continuous or intermittent addition of substrate to a batch reactor or by
feeding of the substrate at several points along a tubular reactor. So far it has been
assumed that the immobilized enzyme obeys the normal enzyme rate equations. However, when substrate diffusion is rate-limiting, the kinetic behavior of the reactor system
is different from that which prevails when the enzyme reaction is rate-determining. Limitation of the reaction rate by the rate of diffusion of substrate from the bulk of the solution to the surface of the immobilized enzyme has been observed in stirred tanks [153]
and in packed beds [154–156]. The productivity of the reactor may change for many
other reasons, some of which are difficult to estimate without carrying out long-term
trials. The immobilized enzyme may be inactivated with time either by denaturation or
by poisoning. Denaturation may result from oxidation, adverse pH or thermal effects.
The choice of the operating temperature is critical. High temperature will increase the
initial output from the reactor and reduce the chance of microbial contamination but
thermal denaturation of enzymes is much higher.

From the above comments, it will be realized that the development of an immobilized
enzyme process is dependent on an overall understanding of enzyme immobilization,
reactor design and operation. Finally, it should be emphasized that many immobilized
enzyme reactors are being operated successfully on both small and large scale. This type
of catalyst system can be expected to take its place among the important types available
for practical application.

All the above considerations have to be considered when deciding upon the most suitable reactor system for starch conversion by means of the appropriate immobilized
enzymes.

6.3 Preparation and Applications of Immobilized α-Amylase Derivatives

It is well known that the microenvironment produced by the carrier may markedly
alter the stability of the bound enzyme. Hydrophylic carriers have proved in many
cases to increase the stability of enzymes covalently bound to such carriers. The binding of α-amylase to different derivatives of cellulose, to Sephadex, to Sepharose and to
other synthetic hydrophylic polymers containing acrylamide residues is described in the
literature [157–169]. Binding was effected in most of the cases via the amino or aro-

matic groups of the enzymes using the corresponding activated polymers. The amount of active enzyme bound was, as a rule, rather small and the high stability of the α-amylase-carrier conjugate was noticed only in a few cases. Because of the rather poor mechanical properties of the carriers so far employed and the practically unsatisfactory stability of the immobilized α-amylases described in the literature, it seems that additional work has to be carried out before testing an adequate preparation on a large scale. Barker et al. [157] prepared water-insoluble derivatives of α-amylase in which the enzyme is covalently bound to microcrystalline cellulose. Coupling was effected via diazotized 3-(p-aminophenoxy)-2-hydroxypropyl ethers and 2-hydroxy-3-(p-isothiocyanatophenoxy) propyl ethers of cellulose possessing different degrees of substitution. The activity retained by the enzyme after coupling amounted only to 2 to 6%. The water insoluble enzyme showed greater heat stability than the native α-amylase, 20% of the original activity remaining after 7 days at 45 °C. The various properties of the cellulose bound α-amylases obtained by Barker et al. [157] were compared by the same authors with immobilized α-amylase preparations obtained by coupling the enzyme with cross-linked copolymers of acrylamide (Enzacryl) [158, 159]. Coupling was effected by diazo isothiocyanate and acid azide groups [158]. The copolyacrylamide derivatives were found to be more stable than the free native enzyme in solution. The most stable α-amylase derivative was that involving Enzacryl A,H, the more hydrophylic carrier. The best preparations thus obtained resemble in their stability the corresponding cellulose based derivatives of α-amylase. With respect to the re-use of the immobilized α-amylases prepared, it was found that the cellulose-bound derivatives lost a considerable amount of physically bound enzyme at the initial stages; subsequent re-use, however, led to comparatively little loss of activity. The Enzacryl-based derivatives were superior in that the activity recovered was good on initial as well as repeated use. Horigome et al. [160] immobilized TAKA-amylase A by covalent binding of the enzyme with Sephadex G-25, Sephadex G-200, Sepharose 6B and Sepharose 2B, activated with cyanogen bromide. The α-amylase-Sepharose 2B conjugate showed the highest activity towards soluble starch (about 7% of that of the soluble amylase). Using a protein staining method, the above authors identified two types of enzyme carrying beads; the Sephadex G-25 enzyme beads containing enzyme on their surface, and the Sephadex and Sepharose-enzyme beads containing enzyme on the surface as well as within the bead. The explanation for this structure of the various enzyme-carrier preparation is given in Fig. 12 [160]. Linko et al. [161] carried out a rather thorough study on the properties and mode of action of B. subtilis α-amylase immobilized by binding with cyanogen bromide activated carboxymethyl cellulose. This carrier was chosen as it was found that α-amylase bound to carboxymethyl cellulose represents a most stable preparation when compared with the other immobilized α-amylase preparations discussed above [162]. The conversion of wheat starch with the immobilized B. subtilis α-amylase was carried out at 72 °C in a stirred tank. The initial reaction rate with the immobilized α-amylase was lower than with the soluble enzyme, but after one hour the immobilized α-amylase produced a higher quantity of reducing sugars than the soluble enzyme (see Figs. 13 and 14). The action pattern of immobilized α-amylase (see Figs. 15, 16) was different from that of the soluble enzyme; immobilized α-amylase produced relatively more glucose and maltose except at the beginning of conversion. This is in accord with the results obtained by

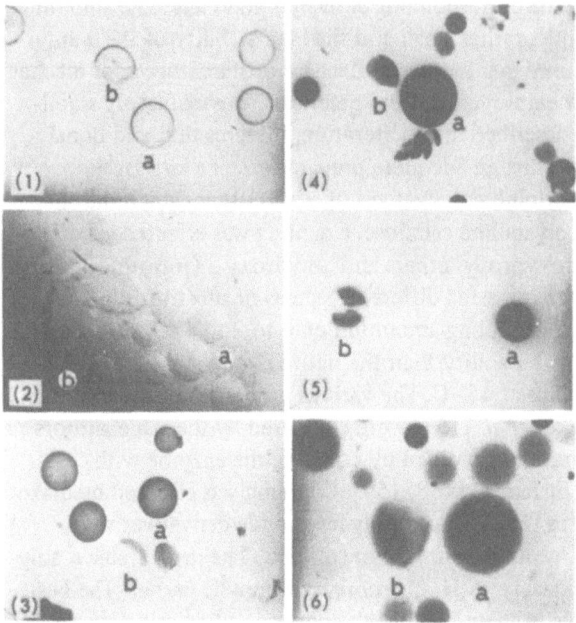

Fig. 12. Photographs of immobilized Taka amylase A stained with nigrosin. Staining was carried out overnight in a 0.25% aqueous solution of nigrosin. Magnification × 100. a, whole spherical gel beads; b, crushed gel beads. (1) Sephadex G-25; (2) Sepharose 2B; (3) Taka amylase A bound to Sephadex G-25; (4) Taka amylase A bound to Sephadex G-200; (5) Taka amylase A bound to Sepharose 6B; (6) Taka amylase bound to Sepharose 2B [160]

Ledingham and Hornby [163] who stated that insoluble derivatives of α-amylase exhibit a higher degree of multiple attack than the native enzyme in solution and a marked increase in the production of glucose. The starch conversion by immobilized α-amylase was not diffusion controlled under the experimental conditions employed. Finally it should be noticed that in spite of the relatively high activity of the α-amylase-cellulose conjugates described above, when used in a batch reactor, they were found to lose a considerable fraction of their activity on re-use (see Table 5) [161].

An interesting study was reported recently by Boundy et al. [164]. The authors observed that immobilization of α-amylase on a phenolformaldehyde resin by adsorption transforms the enzyme activity pattern for starch substrate from an endoenzymic to an exoenzymic mode. In the free state, α-amylase has easy access to the inner recesses of large starch molecules, when initial activity is favored [165]. However, immobilization of the enzyme limits its initial activity to the outermost molecular segments of the polysaccharide. The most striking difference between the action of soluble and thus immobilized α-amylase is the molecular weight distribution of the degradation products as illustrated in Fig. 17. Thus amylose and amylopectin from two starch sources were degraded differently by the native and the immobilized α-amylase. Two distinct fractions were obtained from tapioca amylose using immobilized α-amylase, whereas treatment of tapioca amylose with soluble α-amylase produced a single fraction. Even greater differences were noticed when the degradation products from tapioca amylopectin were compared. These differences in behavior have attributed to steric hindrance and to interaction between the active site of the immobilized enzyme and the amylaceous substrate. Another significant difference between the activity of the soluble and the immobilized enzyme is the rate of hydrolysis of amylose and amylopectin within each

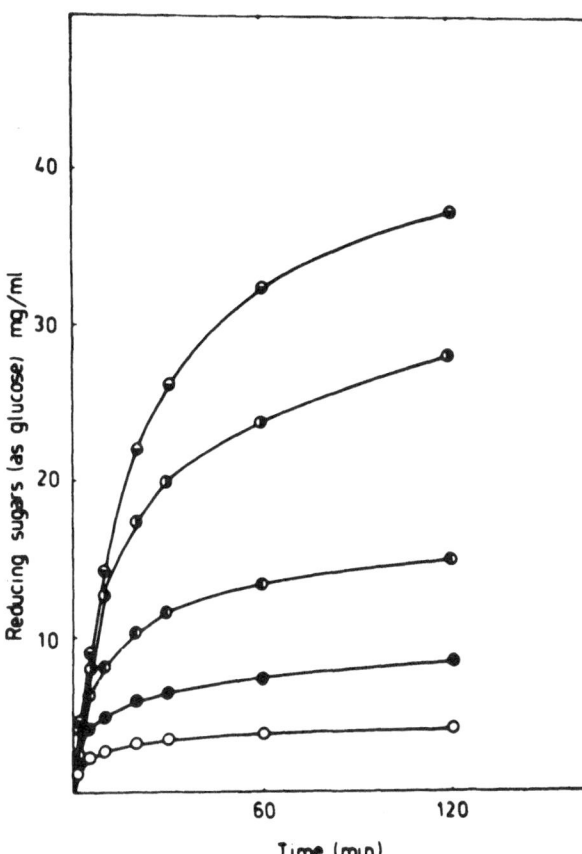

Fig. 13. Starch conversion by
soluble α-amylase (4 300 units/g
starch slurry).
Starch concentrations were:
(○) 1%; (●) 2%; (◐) 4%; (◑) 8%;
(⊖) 12% [161]

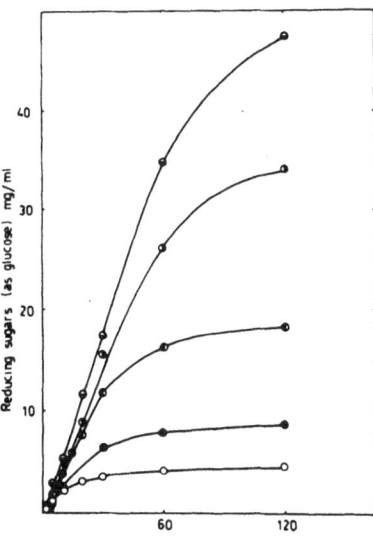

Fig. 14. Starch conversion by immobilized α-amylase (4300 units/g starch
slurry). Starch concentrations were as
in Fig. 13 [161]

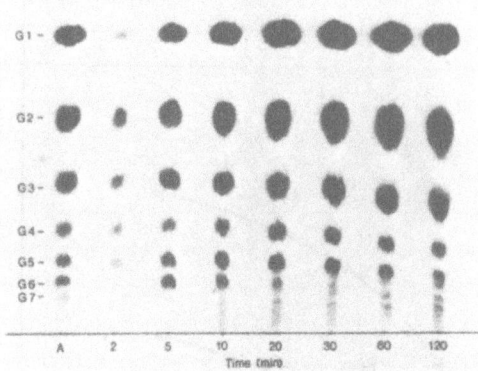

Fig. 15. Chromatographic analysis of immobilized α-amylase (4300 units/g starch slurry) action on starch (1%). The samples were taken at indicated time intervals. Spot A represents reference compounds G_1 through G_7 (oligosaccharides of glucose) [161]

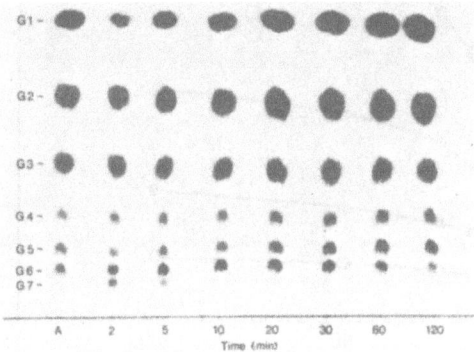

Fig. 16. Chromatographic analysis of soluble α-amylase (4300 units/g starch slurry) action on starch (1%). The samples were taken at indicated time intervals. Spot A represents reference compounds G_1 through G_7 (oligosaccharides of glucose) [161]

Table 5. Recovery of immobilized α-amylase after 1 h conversion of 4% starch slurry[a], (Linko *et al.* [161])

Number of times used	Activity recovered as % of original activity
1	96
2	85
3	74
4	65

[a] Conversion at 72 °C, α-amylase activity 8600 units/g starch slurry.

system, as can be observed from the data in Table 6 [164]. The above results suggest that controlled hydrolysis of starches, particularly cereal starches, with the phenol-formaldehyde immobilized amylases can produce modified starches and oligosaccharides that can readily be separated one from another (see Fig. 17). The modified remaining starch, chiefly amylopectin of high molecular weight might be of use as a non-gelling, thickening agent.

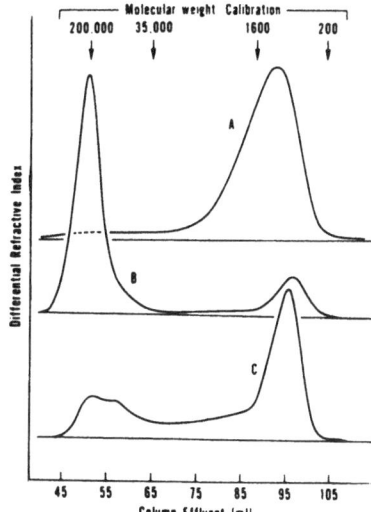

Fig. 17. Molecular weight distribution of tapioca amylase in digestions from soluble and immobilized α-amylase by gel-permeation chromatography (A is a 10-min digestion with soluble enzyme; B and C are 20- and 40-min digestions, respectively, with immobilized enzyme [164])

Table 6. Properties of digests of tapioca-starch fractions with free and bound enzyme (Boundy *et al.* [164])

Substrates	Digest time (min)	Iodine blue color (%[a])	Reducing value (%[b])	Molecular-weight distribution[c] (%)	
				Below 35,000 (d.p.)[d]	Above 35,000
Soluble enzyme					
Amylose	10	77.5	2.8	97 (34.6)	3
Amylopectin	10	51.2	4.9	96	4
Immobilized enzyme					
Amylose	20	79.5	2.7	23 (8.5)	77
	40	30.3	10.7	75 (7.0)	25
Amylopectin	40	89.0	2.0	9	91

[a] Percent of value obtained in unhydrolyzed substrates.
[b] Percent of theoretical value for anhydrous D-glucose.
[c] Determined from relative areas of gel-permeation chromatographic curves.
[d] Degree of polymerization (d. p.) calculated on the assumption that all of the reducing power is present in this fraction.

Finally it is worth mentioning that the binding of α-amylase to complex synthetic hydrophylic copolymers was described in the early sixtees by Manecke and his coworkers [166–168]. The enzyme was coupled to a nitrated copolymer of methacrylic acid, methacrylic acid m-fluoroanilide and divinylbenzene as well as with nitrated copolymers of methacrylic acid 4- or 3-fluorostyrene and divinylbenzene. The enzyme activity of the bound protein in these preparations did not exceed 3% of that of the free enzyme in aqueous solution. The stability of these preparations was reported to be of the same order as that of the free enzyme.

To retain some of the advantages of immobilization, but at the same time, retain good accessibility of high molecular weight substrates, the attachment of enzyme to soluble supports has been investigated [169]. α-Amylase has been immobilized by attachment to soluble amino-S-triazinyl derivatives of dextran 2000, DEAE dextran 2000 and CM-cellulose. These amylase derivatives show up to 67% of the specific activity of the free enzyme, comparing very favorably with the above mentioned insoluble derivatives of α-amylase. The increased stability of soluble immobilized α-amylase with respect to heat compared with free enzyme was demonstrated by the continuous hydrolysis of starch in a ultrafilter reactor (see Fig. 18). About 70% of the activity was retained at 70 °C over a period of 70 h, whereas the free amylase retained 18% of its activity over the same period.

6.4 Immobilized β-Amylase Preparations

β-Amylase has become of great commercial value in the brewing, distilling and baking industries. No wonder, therefore, that various attempts to prepare and use immobilized β-amylase have been made.

A large number of solid supports and various coupling methods have been explored to attain substrate-accessible, highly active and stable preparations. Agarose is one of the most common carriers employed in the immobilization of β-amylase. Among its advantages one should mention large porosity as well as high mechanical stability. Agarose can be prepared in bead form, well suited for column procedure; the enzymic reaction in such a column can be carried out at good flow rates due to the rigidity of the enzyme-beads [170].

Many methods for the coupling of β-amylase to agarose polymers have been described [171, 175–179]. Vretblawd and Axen [171] immobilized *barley* β-amylase by covalent fixation to amino derivatives of epichlorhydrin crosslinked Sepharose, mediated by cyclohexyl isocyanide and acetaldehyde. The enzyme conjugate retains up to 40% of the total activity of the β-amylase added to the coupling mixture. The authors have explored several other methods of binding in order to find a suitable way of attaching β-amylase to agarose. Using cyanogen bromide [172], bis-oxiranes [173] or divinylsulfone [174] under conditions of pH and temperature at which the native enzyme in

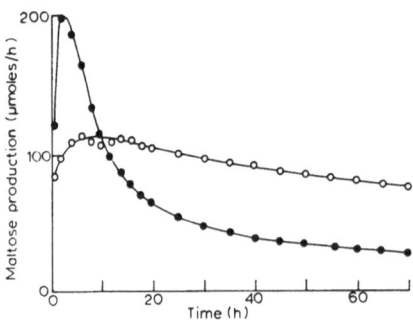

Fig. 18. The continuous hydrolysis of starch in a reactor using amylase (•) and C-M-cellulose-amylase (o) as a catalyst [169]

solution is stable inactive products were obtained. The active β-amylase-Sepharose conjugates prepared showed pH and ionic strength profiles similar to those of native enzyme; however the resistance to inactivation as the result of storage and use was markedly enhanced as a result of immobilization.

The immobilized β-amylases showed remarkable stability when incubated continuously with substrate for prolonged periods of time. Some of the results of experiments performed at different temperatures, using analytical columns containing the gel fixed enzyme, are presented in Fig. 19. At 45 °C half of the initial activity remained after 7 weeks and the corresponding percent at 23 °C was 85%. From the experimental results shown in this figure, it was concluded that 45 °C is the optimal temperature for practical work with the immobilized β-amylase preparations described. Paper chromatography of the hydrolyzed starch after incubation with the immobilized β-amylases showed only maltose in accordance with the action of the soluble enzyme on the same substrate. Interestingly, it was found that an inactivated β-amylase column can be conveniently reactivated by the binding of active native enzyme in the presence of acetaldehyde and cyclohexylisocyanide. This finding might be of considerable practical importance.

Agarose was also the carrier chosen by Caldwell *et al.* [175, 176] for the immobilization of β-amylase by an adsorption procedure. The coupling of β-amylase through hydrophobic interaction between the enzyme and a beaded, crosslinked Sepharose gel, to which hexyl side chains were bonded via ether bridges, was described in detail by the above authors. Hexyl groups have been introduced into crosslinked Sepharose 6B, yielding gels with a degree of substitution in the range of 0.02 to 0.70 mol hexyl-side chains per mole galactose residue. The gels were exposed to β-amylase in solution and the resulting adsorbates showed a monotonic increase in adsorption capacity parallel to the corresponding increase in hexyl content. Absorbate capacity, however, displayed a maximum for a carrier gel with a hexyl-galactose ratio of 0.51. The absolute activity per volume of conjugate compared favorably (40 mg protein/ml gel) with reported activity of β-amylase covalently bound to matrices such as cellulose [177, 178] acrylic polymer [158, 179]

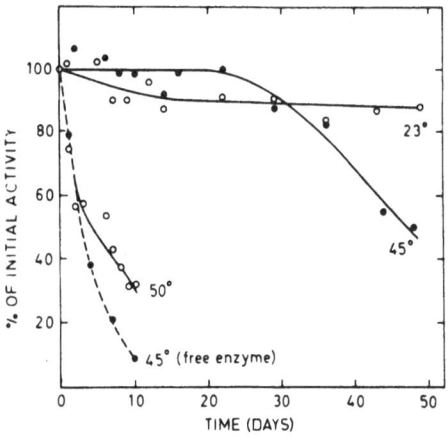

Fig. 19. Continuous digestion of starch by immobilized β-amylase. 0.5 ml enzyme conjugate in a small column (d = 10 mm) was allowed to digest 0.1% starch (in 16 mM acetate, pH 4.8) passing at 10 ml/h. Aliquots of the eluate were assayed. The experiment was carried out at 23, 45 and 50 °C. The dashed line represents the decrease in activity of a solution of β-amylase (15 µg/ml) incubated at 45 °C [171]

and agarose [171]. In spite of the fact that the adsorption procedure described above differs from the immobilization via covalent attachment, the adsorption conjugates might be technically used because of the slow desorption of the enzyme under the regular operational conditions. The preparations obtained were used in column form for continuous maltose production from soluble starch. Special emphasis was given to the three most highly substituted gels because of their potential use in industry. Fig. 20 illustrates the course of activity decay at room temperature for the three enzyme-gel conjugates, continuously percolated with substrate. As can be seen the two adsorbants with a degree of substitution 0.51 and 0.70 showed no evidence of activity release after 1 h, whereas the third conjugate released 3% of its activity in 1 h. It is thus evident that hydrophobic immobilization of β-amylase based on hexyl substituted agarose carrier gel can be optimally performed at a hexyl-galactose ratio ≃ 0.5.

Attempts to immobilize β-amylase by means of diazo or isothiocyanate coupling to cellulose [177] and polyacrylamide beads have also been described. Although β-amylase was readily coupled to the corresponding carriers by all of these procedures, the acid azide-coupled derivative was inactive and the activity retained after diazo and isothiocyanate coupling to polyacrylamide polymers was even lower (~1%) than that of the corresponding cellulose derivatives (~3%). Steric effects, according to the above authors are responsible for the poor activity of the preparations described. The immobilized β-amylases of low activity also showed low stability as compared with the native enzyme.

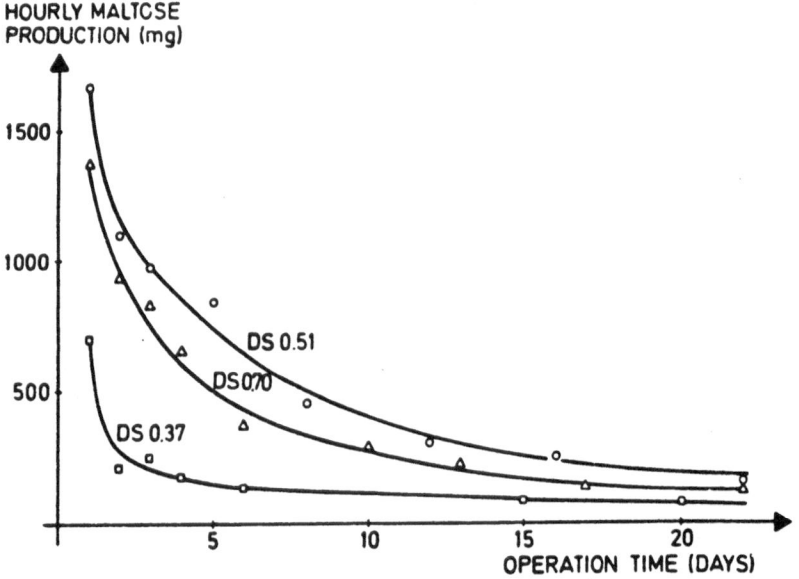

Fig. 20. Operational stability of β-amylase adsorbates with carrier gels of different hexyl-content. (Degree of substitution, D.S., expressed as moles hexyl-group per mole galactose residue.) Adsorbates were formed through incubation of 1 ml suction-dried gel with 2 ml of a β-amylase solution (0.3 mg/ml). 1 ml columns were percolated with a 1% starch substrate (2 ml/h). Activity was assayed against the 5% starch solution. The temperature was 25 °C [176]

An alternative binding procedure to an acrylic copolymer involving carbodiimide was chosen by Märtensson [179]. The product showed 23% β-amylase activity in comparison with that of free enzyme with a coupling yield of 40% based on the amount of added β-amylase.

The operational stability of the free β-amylase and of the different preparations of β-amylase thus immobilized have been investigated by comparing the conversion of starch in a batch treatment using suitable operational conditions (pH 6.0, 45 °C). The results are summarized in Fig. 21. It can be seen that the immobilized preparations were still active after the first 50 h of use and that the residual activity was highest for the glutathione and serum albumin protected preparations. The necessity for immobilized enzymes with high operational stability is quite obvious for successful and economical industrial utilization. Most of the β-amylase immobilized preparations described so far show that β-amylase can be readily immobilized and that some of the immobilized enzymes preparations possess a relatively high stability. The potential industrial use of immobilized β-amylase for continuous production of maltose from starch seems therefore worth considering.

6.5 Preparation and Applications of Immobilized Amyloglucosidase Derivatives

Amyloglucosidase (glucoamylase) is one of the most important industrial enzymes. It is used on a large scale for saccharification of starch to glucose or syrups in the food industry. The process involves the initial thinning of starch by acid or enzyme action and the conversion of the partial hydrolyzed by means of glucoamylase to syrups. The glucoamylase is partly deactivated during the conversion, the remaining active fraction cannot, however, be economically recovered. The replacement of soluble glucoamylase by im-

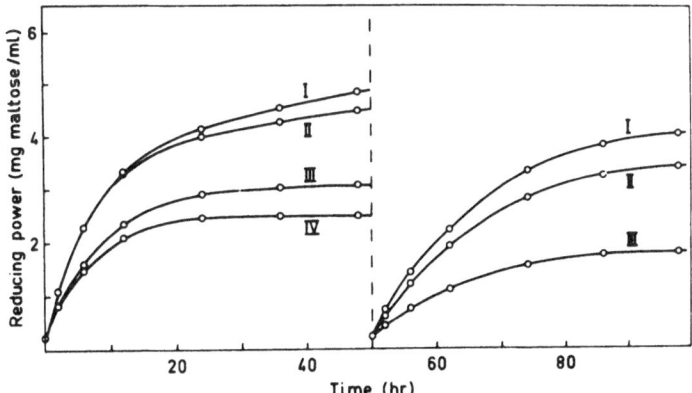

Fig. 21. Operational stability of β-amylase, measured as the hydrolyzing capacity of 1% (w/v) soluble starch in 0.01 M phosphate buffer, pH 6.0 (expressed as mg equivalent of maltose/ml) and 45 °C. The enzyme concentration was 0.83 U/100 ml digest. After 50 h the enzyme was recovered and put into fresh substrate. (I) β-amylase immobilized with red glutathione; (II) β-amylase immobilized with serum albumin; (III) normally immobilized β-amylase; (IV) free soluble β-amylase [179]

mobilized enzyme might allow a more efficient process in which the enzyme can be readily re-used.

More than forty papers, which have been published during the last decade, deal with the immobilization of amyloglucosidase by covalent binding or adsorption to various water-insoluble supports. The properties of the immobilized enzymes have been described and their possible application in the laboratory and in industry discussed. It should be mentioned, however, that as far as this author is aware, no large scale commercial use has been reported, of any of the immobilized amyloglucosidases prepared so far. This stems from the necessity in industry to obtain high conversion, to work at relatively high temperatures (~60 °C) so as to prevent contamination by fungi and bacteria, and to utilize an economic process.

Cellulose and its derivatives have proved to be good support material for enzyme binding in many cases due to their hydrophylic nature, relatively open structure and potentially reactive hydroxyl groups. The first attempt to obtain immobilized amyloglucosidase by binding the enzyme to DEAE-cellulose using the functional reagent 2-amino-4,6-dichloro-S-triazine was described by Wilson and Lilly [180] in 1969. At the same time Maeda and Suzuki [181] reported the preparation of a carboxymethyl-cellulose-glucoamylase complex by coupling the enzyme with the corresponding cellulose azide derivatives. An immobilized glucoamylase preparation has been also described by the Canton group [181], which used p-aminobenzenesulfonylethyl-cellulose as a carrier.

Barker et al. [182] developed a new method of binding amyloglucosidase to cellulose. Inert carriers such as cellulose, nylon or glass were impregnated with titanium chloride, or other metal chlorides, and the thus activated carriers were coupled with the enzyme. The recovered activity of the insoluble preparations amounted to 16–25%; however starch conversion by these enzyme derivatives was considerably lower than those required by industry. In spite of the relative stability of the amyloglucosidase preparations covalently bound to the cellulose as described above, practically all of them were found to lose a marked fraction of their catalytic activity upon re-use or continuous application.

The gel entrapment method—enzyme being physically entrapped in the three dimensional matrix of a gel—has been chosen by many authors because of its simplicity and convenience. The method can be applied to any enzyme and no reactive groups of the enzyme are required for attachment as in the covalent immobilization techniques. Glucoamylase has been encapsulated within crosslinked poly(acrylamide) gels and studied in some detail by Gruesback and Rase [183]. The resulting complex retained 22% of the activity of the free enzyme. The entrapped enzyme was much more stable with respect to temperature than the free enzyme. Unfortunately the gel hindered the penetration by the catalyst of large substrate molecules. Beck and Rase [184] recommend encapsulating the enzyme in polyacrylamide gel by suspension rather than bulk polymerization, followed by reactivation of the already entrapped enzyme by sulfhydryl treatment. By this procedure, immobilized preparations with an activity of 98% of the original were obtained.

Walton et al. [185] used immobilized glucoamylase prepared by gel entrapment to continuously convert partially hydrolyzed starch to 90–94% glucose syrups. The hydrolysis power of their gel for various dextrins was not determined by the above authors. Entrapment of glucoamylase was also carried out by using N-vinylpyrolidone (VP) monomer [186] and initiating polymerization by X-ray irradiation. Forty-five percent of the enzymic activity was recovered in V. P. glucoamylase gel. The temperature and pH stability of the immobilized enzyme were inferior to those of the native enzyme. The Japanese group also [186] evaluated the hydrolysis power of its gels towards various dextrins. It was found that V. P. glucoamylase gel hydrolyzes a high molecular weight dextrin (mol. wt. 10,400) almost completely and that the glucose equivalent obtained was equal to that recorded for the native enzyme. These results suggest that an entrapped enzyme of the type described might be of practical use.

The binding of enzymes to adequate carriers by adsorption undoubtedly represents the most simple and economical procedure for enzyme immobilization Smiley et al. [187] have obtained DEAE-cellulose-bound enzyme derivatives by physical, partially ionic adsorption. The activity of the immobilized amyloglucosidase ranged from 16 to 55% of the activity of the free enzyme. A batch-type procedure was used to establish the optimum conditions for testing the activity of bound enzyme. As DEAE-cellulose is a compact support material and the enzyme is adsorbed mainly by electrostatic forces, it is not tightly bound and difficulties were encountered in practical application.

Adsorption of amyloglucosidase to a positively charged resin can be improved by increasing the negative charge on the enzyme [188]. Highly charged water soluble conjugates of amyloglucosidase with a copolymer of ethylene-maleic acid or styrene-maleic acid were prepared and adsorbed on DEAE-cellulose and other cationic resins. Negatively charged enzyme conjugates were also obtained by succinylation. The polyanionic conjugates of amyloglucosidase adsorbed on the cationic carriers showed improved temperature stability over the native enzyme adsorbed on the same carriers under the same experimental conditions. Retentional activity (up to 40%) was improved as well as mechanical properties and operational stability (see Table 7) [188]. The insoluble enzyme conjugate-carrier complexes could be re-used and retained full activity while working continuously for 3 weeks.

The Chinese group [189] recommended DEAE-Sephadex as the most suitable carrier for the adsorption of amyloglucosidase. The relative retentional activity, calculated on the basis of ability to saccharify starch reached 35–45%. After being used ten times under optimal conditions, the immobilized enzyme still retained 63% of its original activity. Recently Caldwell et al. [190] immobilized glucoamylase by adsorption onto a hexyl-

Table 7. Stability of immobilized amyloglucosidase at 50 $^\circ$C in buffer acetate, 0.1 M, pH 4.2 (Solomon and Levin [188])

Enzyme preparations	Residual activity[a] (%)		
	1 day	2 days	3 days
AG	89	53	18
EMA-AG-DEAE-cellulose	71	57	50
AG-DEAE-cellulose	69	46	29

[a] Activity determined on NPG.

Sepharose; retentional activity upon immobilization was high, varying from full activity at low enzyme content to 68% at the adsorption limit of the carrier. Continuous operation for three months reduced the activity of the conjugate to 40%. The thermal stability of the adsorbate was inferior to that of soluble enzyme, but was noticeably enhanced in the presence of substrate.

Amyloglucosidase has been adsorbed on inorganic carriers [191] such as acid-activated molecular sieve and alumina. The immobilized enzyme preparations exhibited 50–100% of the initial enzyme activity and possessed high temperature stability. A prolonged working life span was reported; some of the preparations retained full activity for up to three weeks while drops in activity were recorded after five weeks. The good mechanical and flow characteristics suggest possible industrial application.

Inorganic materials including ceramics, glass and metal have also been used for the covalent attachment of amyloglucosidase. For example amyloglucosidase was coupled to arylamine and alkylamine glass with a coupling efficiency average of 40–80% [154]. By making use of enzyme columns of the glass-amyloglucosidase it was shown that the decay of activity was temperature and flow-rate dependent. It was pointed out that the loss of activity of an immobilized enzyme in a column reactor may occur in several ways. e.g., by enzyme denaturation, by leakage of small enzyme-carrier granules, etc. It was also of interest to determine whether the decay in activity was due to the rupture of any enzyme carrier bond during the flow of substrate. Weetall *et al.* [192] calculated the shear forces necessary to tear the enzyme from the preparate. The calculation of the prevailing shearing forces, based on the assumption of a single covalent bond, shows that the energy required to shear one covalent bond is at least ten orders of magnitude greater than that generated in a fixed bed enzyme column under the usual experimental conditions. The possibility of losing enzyme by rupture of the covalent linkage between enzyme and carrier seems, therefore, extremely unlikely and can be eliminated as a possible mechanism for loss of enzymic activity. Weetall [193] examined the possible use in industry of the amyloglucosidase-glass system in great detail. He defined the operating parameters required for a scale up of the system using real conditions of feed concentration, 30% solids and temperatures of 40°–65 °C. Based on these studies, Weetall [193] worked out a plan for a pilot plant capable of producing 5 000 t of dextrose per year. The parameters of the plant are given in Table 8, and it was claimed that

Table 8. Parameters for dextrose production pilot plant for the production of 5000 t of dextrose per year (Weetall [193])

Enzyme:	Partially purified amyloglucosidase
Substrate:	30% (dry substance) enzyme thinned cornstarch
Carrier:	Silanized inorganic support
Operating pH:	4.5
Temperature:	40 °C–50 °C
Reactor size:	0.1–0.2 m^3
Specific activity:	3000 units/g derivative (1 unit produces 13.8 mg dextrose per hour at 60 °C)
K_m (app):	3×10^{-4} M
Reactor type:	Plug flow

the system proposed is economically attractive. Another interesting aspect concerns substrate residence time. Normally the time required for the production of 95 D.E. corn syrup using native enzyme in a batch process is approximately 75 h. However, in the pilot plant proposed the time required is less than 60 min. The half-life of the enzyme can be increased substantially by operating at lower temperatures without greatly affecting the reaction rate. As can be seen from Table 9 [193], by operating at 40 to 45 °C the half life is extended over a period of two to three years with only a 70 to 80% decrease in reaction rate as compared with the operation rate at 60 °C. The system recommended is small, compact, inexpensive and completely automated.

A systematic study has been made to determine the controlling mass transfer resistance on the overall reaction rate in the conversion of maltose to glucose, catalyzed by gluco-amylase immobilized onto porous glass [194]. For normal operation of a packed column or an air-stirred batch reactor, the rate controlling step was found to be the internal resistance of simultaneous pore diffusion and chemical reaction. The temperature-enzyme activity profile of the glucoamylase derivatives using maltose as the substrate is shown in Fig. 22. The temperature optimum of the glucoamylase bound on porous glass is about 60 °C. If the feed to an adiabatic glucoamylase fixed bed reactor was at 60 °C, a temper-ature jump of even 10 °C could reduce the enzyme activity to about 55% of the maxi-mum. The temperature effect raises a practical problem of reactor operation and it has been shown [195] that the temperature optimum on a temperature-enzyme activity profile may not necessarily be the most desirable input temperature to a large, fixed bed enzyme reactor. In the case of immobilized glucoamylase it was found more effective to feed the reactor slightly below the optimum temperature and let the heat of reaction warm the reaction mixture to achieve the maximum overall conversion through the reactor. It was calculated that the most desirable feed temperature to an adiabatic gluco-amylase-glass column is 58.5 °C rather than the optimum temperature (60 °C) of the enzyme activity-temperature profile in Fig. 22. The procedure of Marsh, based on gluco-amylase attached to alkylamine porous silica glass using glutaraldehyde [195], was employed by Lee et al. [196] in designing a pilot plant for the production of glucose (see Fig. 23). For this purpose the kinetics and stability of the immobilized enzyme were studied with acid and α-amylase hydrolyzed dextrin as the substrate. The enzyme was found to be extremely stable in both laboratory and pilot plant operations. The

Table 9. Half-life and rate temperature dependence of amyloglucosidase-glass system (Weetall [193])

Temperature °C	Half-life days	Relative reaction rate (% of that at 60 °C)	Kilograms dextrose[a] per gram carrier
60	13	100	5.3
50	100	70	18.9
45	645 (extrapolated)	30	51.9
40	900 (extrapolated)	25	61.0

[a] Approximation assuming initial activity of immobilized glucoamylase is 2000 IU/g derivative.

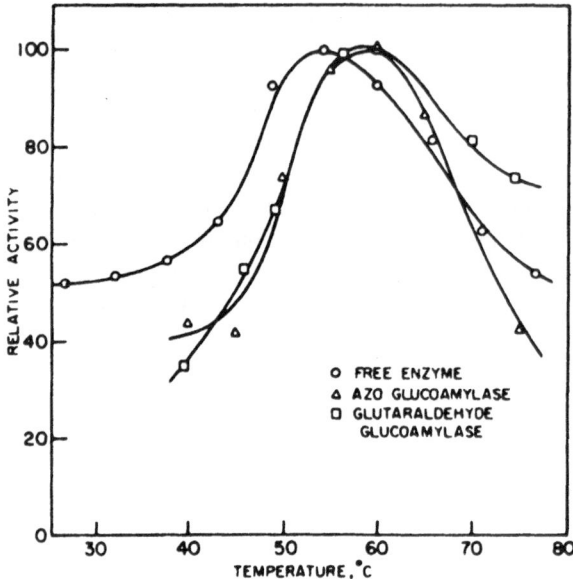

Fig. 22. Temperature-enzyme activity profile of glucoamylase [195]

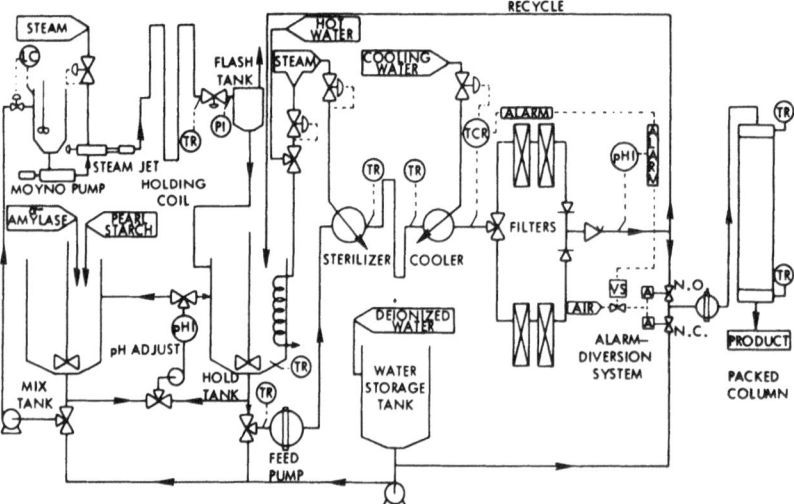

Fig. 23. Process flow diagram of immobilized glucoamylase pilot plant [196]

pilot plant glucoamylase column was operated continuously for 80 days at 38 °C with 30% acid thinner dextrin feed as substrate. During this period there was no loss of activity. It was also demonstrated that the type of feed employed with immobilized glucoamylase has a marked effect on the final yield of glucose. When the feed had been previously only lightly hydrolyzed, pore diffusion limitation caused an appreciable decrease in the glucose production rate. The diffusional gradients present in the carrier pores

caused the immobilized enzyme to yield lower glucose concentrations than the free
enzyme at similar feed conditions.

Attempts to minimize the importance of diffusional resistance within the support have
been reported by the French group [197]. For this purpose, amyloglucosidase was covalently bound to collagen sheets via acyl azide activation of the support. Comparative
kinetic studies using maltose showed that the optimum pH and temperature, activation
energy and K_m of the free enzyme were not affected after grafting to the collagen.
Although the K_m was increased fivefold with soluble starch after grafting, the rate of
hydrolysis of this substrate was still four times that for maltose. Thus, external diffusional resistance appeared negligible for maltose, but significant for soluble starch. The
ratio V_m (soluble starch) V_m (maltose) for free amyloglucosidase was maintained after
grafting amyloglucosidase onto collagen. Another advantage of enzymic membranes of
collagen is the striking resistance to bacterial contamination probably due to the acyl
azide activation of the support before grafting. Their stability with respect to storage
and under working conditions was very good, since the enzyme-membranes kept full
activity after storage at 4 °C for 9 months and 80% of their activity after 17 months of
storage. Continuous operation for 18 days at 40 °C in a helicoidal reactor did not affect
the activity of the bound glucoamylase.

Various other methods of preparation of immobilized glucoamylase have been reported
[198–202], unfortunately the utilization of these preparations seems, doubtful. In spite
of the considerable number of studies dealing with immobilized amyloglucosidase, there
is still much to be learned about the preparation, the factors determining enzyme stability and the use of the enzyme in reactor systems. Under practical operating conditions
it is difficult to attain the same degree of conversion of prehydrolyzed starch to dextrose
as can be achieved with soluble glucoamylase under similar conversion conditions. The
interplay of hydrolytic and transferase reactions, leading to isomaltose in immobilized
systems in which the concentration of the enzyme is very high, is probably of significance and is rather different from that to be found in batch operations using soluble
enzyme, where relatively low concentrations of enzyme are present throughout the reaction period.

Further research on the fundamental aspects of the mechanism of reactions catalyzed
by immobilized glucoamylase may generate new concepts that will finally provide the
basic information required for the manufacture of crystalline dextrose.

6.6 Immobilized Pullulanase

Debranching enzymes such as pullulanase have many potential industrial applications.
Examples in the starch industry [203] include the production of low molecular weight
amylose and high purity maltose, their use as effective aids in the preparation of stabilized starch solutions of low viscosity as well as for the production of maltotriose in
high yield from pullulan. Pullulanase [204] also appears to be an alternative carbohy-
drase for use in brewing from unmalted cereals. Mindful of the potential of this enzyme
and of the advantages of immobilized enzymes, Mårtensson and Mosbach [205] covalently bound pullulanase to an inert crosslinked copolymer of acrylamide-acrylic acid
by using a water soluble carbodiimide. The binding yield based on the amount of added

pullulanase was 34%. Coupling in the presence of pullulan gave a 5-fold increase in activity over that obtained when the substrate was absent. However, no increase in stabilization above that of the soluble enzyme was obtained. The immobilized pullulanase was used in a packed bed column to debranch amylopectin to low molecular weight amylose. Use of insoluble pullulanase in a large scale process will require further studies aimed in particular at stabilization of the immobilized form of the enzyme.

One application of immobilized pullulanase of particular interest to the starch industry is the debranching of amylopectin to low molecular weight amylose, which in a two-step enzymic reaction could subsequently be hydrolyzed to maltose using immobilized β-amylase.

6.7 Immobilized Two Enzyme System: β-Amylase and Pullulanase

Since immobilized preparations of β-amylase [179] and pullulanase [205] had been prepared by covalent coupling to an inert acrylic matrix (Bio-Gel GM 100) using the carbodiimide method, it was logical to design a coupling procedure for the two enzymes to the same matrix surface [206]. Two-enzyme systems have been previously described [207, 208] in which the enzymes operate in sequence, the product of the first enzyme serving as a substrate for the second. Here another variant was presented where a steric opening of the substrate is performed by one enzyme, thus facilitating the action of the other enzyme. Furthermore, coupling of the enzymes to the same matrix surface was preferred, rather than the use of a mixture of the two individually immobilized preparations, in view of the greater stability of immobilized β-amylase at a high local protein concentration, as described elsewhere [205]. The coupling was performed in two successive steps because of the different optimal coupling conditions of the two enzymes. The coupling yields of β-amylase protein and pullulanase protein were 40% and 38%, respectively. The enzymes showed a residual catalytic activity of 22% and 32%, respectively.

In order to find the optimal operational conditions, the action of the immobilized two-enzyme system on partially hydrolyzed starch was studied using a packed column. It was found that pH 6.0 and 45 °C were the most prefereable conditions for long-time runs. A stability test (Fig. 24) showed that after 50 h the preparation was still active, whereas the free enzymes lose practically all of their activity after 20 h of operation. The decrease in the degree of conversion over a period of four weeks at a constant flow rate is shown in Fig. 25. The use of the two enzymes in the starch industry offers an interesting possibility for the complete conversion of starch. The conversion of starch into a product of high maltose content, using a mixture of soluble β-amylase and pullulanase had been described previously [209–210]. Märtensson [211] has reported kinetic studies on the immobilized two-enzymes system of β-amylase and pullulanase by comparison between the theoretical and practical efficiencies of two of the basic types of continuous enzyme reactors (CSTR and PFR). The process kinetics for the conversion of starch to a high maltose containing product with the aid of the immobilized two-enzymes seem to favor the use of a PFR compared to a CSTR reactor.

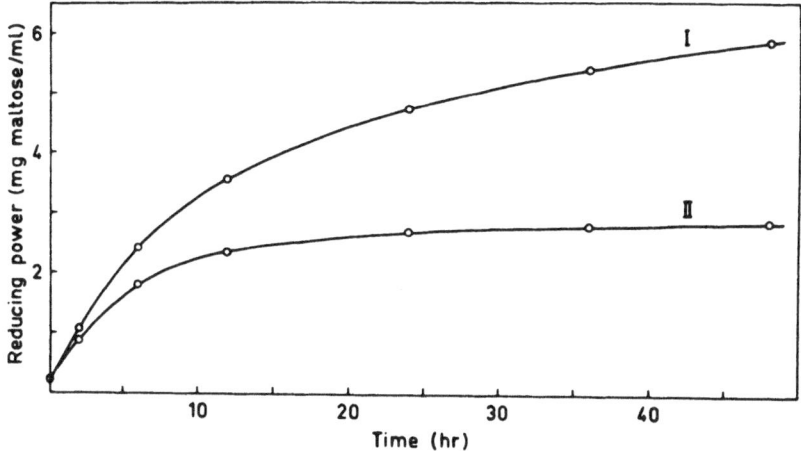

Fig. 24. Operational stability of β-amylase-pullulanase, measured as the hydrolyzing capacity of 1% soluble starch (expressed as mg equivalent of maltose) in 0.01 M phosphate buffer, pH 6.0, at 45 °C; β-amylase-pullanase concentrations were 0.65 U and 0.098 U, respectively, per 100 ml digest: (I) immobilized β-amylase-pullulanase preparations; (II) free soluble enzymes [206]

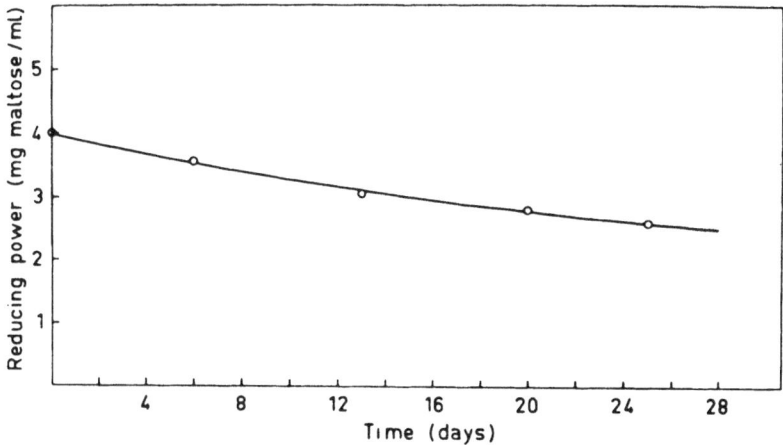

Fig. 25. Operational stability of immobilized β-amylase-pullulanase (14.5 U and 2.2 U, respectively) used in a packed bed with a height of 7 cm (column I). Substrate 0.5% (w/v) starch DE 3.4 in 0.01 M phosphate buffer, pH 6.0 at 45 °C. Flow rate: 13.8 ml/h. Degree of conversion is expressed as mg equivalent of maltose/ml [207]

7 Concluding Remarks

The findings summarized show clearly that enzyme science and technology are of considerable importance in the development of important new techniques involving the use of enzymes in the production of modified starches; i.e. glucose, maltose and fructose containing syrups as well as crystalline dextrose.

Immobilization techniques offering an attractive opportunity for the multiple usage of the same enzyme has made enzymology a feasible approach for the production of new derivatives on an industrial scale and has introduced a new field namely Enzyme Engineering into process development.

It is worth noting that the controlled use of enzymes in the hydrolysis of starch and of immobilized enzymes, which can be readily recovered and re-used, show considerable promise. The development of "Enzyme Engineering", when applied to starch conversion, lies in the need to transform the information already available and the potential applications into actual practise. In this case, as well as in many others, it has still to be proved that the new enzyme technologies worked out in the laboratory are practical, economical and more efficient than the well established industrial processes.

8 Acknowledgement

I am very grateful to Professor Ephraim Katchalski-Katzir for his help and encouragement in writing this manuscript.

9 References

1. Kirchoff, G. S. C.: Acad. Imp. Sci. St. Petersbourg, Mem. **4**, 27 (1811)
2. de Saussure, T.: Bull. Pharm. **6**, 499 (1814)
3. Biot, J. B., Persoz, J.: Ann. Chim. Phys. **52**, 72 (1833)
4. Bondonneau, L.: Comp. Rend. **81**, 972 (1875)
5. Salomon, F.: J. Prakt. Chem. **25**, 348 (1882)
6. Schulze, L.: J. Prakt. Chem. **28**, 311 (1883)
7. Meyer, A.: Untersuchungen über die Stärkekörner. Yena: G. Fisher, 1896
8. Reichert, E. T.: Differentiation and Specificity of Starches. Washington: Carnegie Publ., 1913
9. Walton, R. P.: A Comprehensive Survey of Starch Chemistry. New York: J. Reihold, 1928
10. Karrer, P.: Helv. Chim. Acta **3**, 620 (1920)
11. Karrer, P., Nägeli, C.: Helv. Chim. Acta **4**, 185 (1921)
12. Irvine, J. C., MacDonald, J.: J. Chem. Soc. **1926**, 1502
13. Haworth, W. N., Hirst, E. L., Webb, J. J.: J. Chem. Soc. **1928**, 2681
14. Staudinger, H.: Die Hochmolekularen Organischen Verbindungen, Kautschuk und Cellulose. Berlin: Springer-Verlag, 1932
15. Carothers, W. H.: Collected Papers of W. H. Carothers on High Polymers. New York: Interscience Publishers Inc., 1940
16. Meyer, K. H.: Starch and Related Carbohydrates. In: Natural and Synthetic High Polymers. Mark, H., Kraemer, E. O., Whitby, G. J. (eds.). New York: Interscience Publishers Inc., 1942, pp. 387–422
17. Whistler, R. L.: Starch–its Past and Future. In: Starch: Chemistry and Technology, Vol. I., Whistler, R. L., Paschall, E. F. (eds.). New York and London: Academic Press, 1965, pp. 1–8
18. Sandstedt, R. M.: Cereal Chem. Suppl. **32**, 17 (1955)
19. Badenhuizen, N. P.: Occurrence and Development of Starch in Plants. In: Starch: Chemistry and Technology, Vol. I. Whistler, R. L., Paschall, E. F. (eds.). New York and London: Academic Press, 1965, pp. 65–100

20. Greenwood, C. T.:, Thomson, J.: J. Chem. Soc. **222** (1962)
21. Greenwood, C. T.: Starch and Glycogen. In: The Carbohydrates-Chemistry and Biochemistry. Pigman, W., Horton, D. (eds.), Vol. II B. New York, London: Academic Press, 1970, pp. 471–513
22. Adkins, G. K., Greenwood, C. T.: Stärke **18**, 213 (1966)
23. Wolfrom, M. L., El Khadem, H. S.: Chemical Evidence for the Structure of Starch. In: Starch: Chemistry and Technology, Vol. I. Whistler, R. L., Paschall, E. F. (eds.) 1965, pp. 251–274
24. Pazur, J. H.: Enzymes in Synthesis and Hydrolysis of Starch. In: Starch-Chemistry and Technology. Whistler, R. L., Paschall, E. F. (eds.). Vol. I. New York, London: Academic Press, 1965, pp. 155–171
25. Schoch, T. J.: Advan. Carbohyd. Chem. **1**, 247 (1945)
26. Whistler, R. L., Johnson, C.: Cereal Chem. **25**, 418 (1948)
27. Stein, R. S., Rundle, R. E.: J. Chem. Phys. **16**, 195 (1948)
28. Murakami, H.: J. Chem. Phys. **22**, 367 (1954)
29. Bersohn, R., Isenberg, I.: J. Chem. Phys. **35**, 1640 (1961)
30. Gilbert, G. A., Marriott: J. V. R. Trans. Faraday Soc. **44**, 84 (1948)
31. Bates, F. L., French, D., Rundle, R. E.: J. Amer. Chem. Soc. **65**, 142 (1943)
32. Anderson, D. M. W., Greenwood, C. T.: J. Chem. Soc. **1955**, 3016
33. Adkins, G. K., Greenwood, C. T.: Carbohyd. Res. **3**, 81,152 (1966)
34. Whelan, W. J.: Starch and Similar Polysaccharides. In: Encyclopaedia of Plant Physiology. Ruhland, W. (ed.) Vol. 6. Berlin: Springer-Verlag, 1958, pp. 154–240
35. Whistler, R. L., Paschall, E. F.: Fundamental Aspects. In: Starch: Chemistry and Technology, Vol. I. New York: Academic Press, 1965
36. Radley, J. A.: Starch and its Derivatives, 4th ed., London: Chapman and Hall, 1968
37. Greenwood, C. T.: Stärke **12**, 169 (1960)
38. Abdullah, M., Lee, E. Y. C., Whelan, W. J.: Biochem. J. **97**, 10P (1965)
39. Lee, E. Y. C., Whelan, W. J.: Arch. Biochem. Biophys. **116**, 162 (1966)
40. Adkins, G. K., Banks, W., Greenwood, C. T.: Carbohyd. Res. **2**, 502 (1966)
41. Haworth, W. N., Hirst, E. L., Isherwood, F. A.: J. Chem. Soc. **1937**, 577
42. Staudinger, H., Husemann, E.: Ann **527**, 195 (1937)
43. Meyer, K: H., Bernfeld, P.: Helv. Chim. Acta **23**, 875 (1940)
44. Myrbäck, K., Sillen, L. G.: Acta Chem. Scand. **3**, 190 (1949)
45. Peat, S., Whelan, W. J., Thomas, G. J.: J. Chem. Soc. **1952**, 4546
46. Graeza, R.: Minor Constituents of Starch. In: Starch: Chemistry and Technology. Whistler, R. L., Paschall, E. F. (eds.), Vol. I. New York-London: Academic Press, 1965, pp. 105–128
47. Schoch, T. J.: J. Am. Chem. Soc. **64**, 2954 (1942)
48. Pasternak, T.: Helv. Chim. Acta **18**, 1351 (1935)
49. Lampitt, L. M., Fuller, C. H. F., Goldenberg, N.: J. Soc. Chem. Ind. (London) **67**, 121 (1948)
50. Hodge, J. E., Montgomery, E. M., Hilbert, G. E.: Cereal Chem. **25**, 19 (1948)
51. Radowsky, M. W., Smith, M. D.: Cereal Chem. **40**, 31 (1963)
52. Taylor, T. C., Nelson, T. H.: J. Am. Chem. Soc. **42**, 1726 (1920)
53. Lehrman, L.: J. Am. Chem. Soc. **61**, 212 (1939)
54. Reed, G. (ed.): Enzymes in Food Processing. New York, San Francisco, London: Academic Press, 1975
55. Lee, E. Y. C., Whelan, W. J.: In: The Enzymes. Boyer, P. D., (ed.) Vol. 5. New York: Academic Press, 1971, pp. 191–234
56. Fischer, E. H., Stein, E. A.: Arch. Sci. **7**, 131 (1954)
57. Fischer, E. H., Stein, E. A.: Biochem. Prep. **8**, 27 (1961)
58. Schwimmer, S., Balls, A. K.: J. Biol. Chem. **179**, 1063 (1949)
59. Fischer, E. H., Stein, E. A.: In: The Enzymes, Boyer, P. D., Lardy, H. A., Myrback, K., (eds.) 2nd ed. Vol. 4. New York: Academic Press, 1960, pp. 313–342
60. Whitaker, J. R.: In: Principles of Enzymology for the Food Sciences. New York: Dekker, 1972, pp. 433–467
61. Walker, G. J., Whelan, W. J.: Biochem. J. **76**, 257 (1960)

62. Freeman, G. G., Hopkins, R. H.: Biochem. J. **30**, 442 (1932)
63. Giri, K. V., Nigam, V. N., Saroja, K.: Naturwissenschaften **40**, 484 (1953)
64. Pazur, J. H., French, D., Knapp, D. W.: Proc. Iowa Acad. Sci. **57**, 203 (1950)
65. Pazur, J. H., Budovitz, T.: J. Biol. Chem.: **220**, 25 (1956)
66. Pazur, J. H., Sandstedt, R. M.: Cereal Chem. **31**, 416 (1954)
67. Manning, G. B., Campbell, L. L.: J. Biol. Chem. **236**, 2952 (1961)
68. Junge, J. M., Stein, E. A., Neurath, H., Fischer, E. H.: J. Biol. Chem. **234**, 566 (1959)
69. Meyer, K. H., Fuld, M., Bernfeld, P.: Experientia **3**, 411 (1947)
70. Akabori, S., Ikemura, T., Hagihara, B.: J. Biochem. (Tokyo) **41**, 577 (1954)
71. Fischer, E. H., Montmollin, R.: Helv. Chim. Acta **34**, 1994 (1951)
72. Mayer, F. C., Larner, J.: J. Am. Chem. Soc. **81**, 188 (1959)
73. Koshland, D. E., Jr.: In: The Mechanism of Enzyme Action. McElroy, M. D., Glass, B. (eds.). Baltimore, Maryland: John Hopkins Press, 1954, p. 608
74. Balls, A. K., Thompson, R. R., Walden, M. K.: J. Biol. Chem. **163**, 571 (1946)
75. French, D.: In: The Enzymes. Boyer, P. D., Lardy, H., Myrback, K. (eds.), Vol. 4. New York: Academic Press Inc., 1960, pp. 345–368
76. England, S., Singer, T. P.: J. Biol. Chem. **187**, 213 (1950)
77. Meyer, K. H., Spahr, P. F., Fischer, E. H.: Helv. Chim. Acta **36**, 1924 (1953)
78. Piguett, A., Fischer, E. H.: Helv. Chim. Acta **35**, 257 (1952)
79. Balls, A. K., Walden, M. K., Thompson, R. R.: J. Biol. Chem. **173**, 9 (1948)
80. Fukumoto, J., Tsujisaka, Y.: Kagaku Kogyo **28**, 282 (1954)
81. Gertler, A., Birk, Y.: Biochim. Biophys. Acta **118**, 98 (1966)
82. Visuri, K., Nummi, M.: Eur. J. Biochem. **28**, 555 (1972)
83. Thoma, J. A., Koshland, D. E.: J. Mol. Biol. **2**, 169 (1960)
84. Myrbäck, K.: Ark. Kemi **2**, 417 (1950)
85. Meyer, K. H.: Experientia **8**, 405 (1952)
86. Summer, R., French, D.: J. Biol. Chem. **222**, 297 (1960)
87. Cleveland, F. C., Kerr, R. W.: Cereal Chem. **25**, 133 (1948)
88. Hopkins, R. H., Jelinek, B., Harisson, L. E.: Biochem. J. **43**, 32 (1948)
89. England, S., Sorof, S., Singer, T. P.: J. Biol. Chem. **189**, 217 (1951)
90. Bailey, J. M., French, D.: J. Biol. Chem. **226**, 1 (1957)
91. Bailey, J. M., Whelan, W. J.: Biochem. J. **67**, 540 (1957)
92. Husemann, E., Burchard, W., Pfannemüller, B.: Stärke **16**, 143 (1964)
93. Kitahara, K., Kurushima, M.: Makko Kogatu Zasshi **27**, 254 (1949)
94. Kerr, R. W., Cleveland, F. C., Katzbeck, W. J.: J. Am. Chem. Soc. **73**, 3916 (1951)
95. Corman, J., Langlykke, A. F.: Cereal Chem. **25**, 190 (1948)
96. Phillips, L. L., Caldwell, M. L.: J. Am. Chem. Soc.: **73**, 3559 (1951)
97. Hopkins, R. H., Kelka, D.: Arch. Biochem. Biophys. **69**, 45 (1957)
98. French, D., Knapp, D. W.: J. Biol. Chem. **187**, 463 (1950)
99. Hayashida, S.: Bull. Agr. Chem. Soc., Japan **21**, 386 (1957)
100. Tsujisaka, Y., Fukumoto, J., Yamamoto: Nature **181**, 770 (1958)
101. Freedberg, I. M., Levin, Y., Kay, C. M., McCubbin, W. D., Katchalski-Katzir, E.: Biochim. Biophys. Acta **391**, 361 (1975)
102. Pazur, J. H., Ando, T.: J. Biol. Chem. **235**, 297 (1960)
103. Pazur, J. H., Kleppe, K.: J. Biol. Chem. **237**, 1002 (1962)
104. Pazur, J. H., Ando, T.: J. Biol. Chem. **234**, 1966 (1959)
105. Phillips, L. L., Caldwell, M. L.: J. Am. Chem. Soc. **73**, 3563 (1951)
106. Burgher, M., Beran, K.: Collection Czech. Chem. Commun. **22**, 299 (1957)
107. Weill, C. E., Burch, R. J., Van Dyck, J. M.: Cereal Chem. **31**, 150 (1954)
108. Lee, E. Y. C., Carter, J. H., Nielsen, L. D., Fischer, E. H.: Biochemistry **9**, 2347 (1970)
109. Manners, D. J.: Nature New Biology (London) **234**, 150 (1971)
110. Hobson, P. N., Whelan, W. J., Peat, S.: J. Chem. Soc. **1951**, 1541
111. Lee, E. Y. C., Marshall, J. J., Whelan, W. J.: Arch. Biochem. Biophys. **143**, 365 (1971)
112. Marshall, J. J.: Wallerstein Lab. Commun. **35**, 49 (1972)

113. Whelan, W. J., Roberts, P. J. P.: Nature (London) **170**, 748 (1952)
114. Drummond, G. S.: Diss. Abstr. B. **31**, 1696 (1970)
115. Marada, T., Yokobayashi, K., Misaki, A.: Appl. Microbiol. **16**, 1439 (1968)
116. Gunja-Smith, Z., Marshall, J. J., Smith, E. E., Whelan, W. J.: FEBS Lett. **12**, 96 (1970)
117. Maruo, B., Kobayashi, T.: Nature (London) **167**, 606 (1951)
118. Lee, E. Y. C., Smith, E. E., Whelan, W. J.: In: Miami Winter Symposia. Whelan, W. J., Schultz, J. (eds.), Vol. 1. Amsterdam: North Holland Publ., 1970, p. 139
119. Rubin, D. H.: Technology Assessment Series: Enzymes (Industrial) National Technical Information Service, PB 202778-04, Springfield, Va. 1971
120. Abdullah, M.: Westmont, Ill. U.S. Patent 3654082 (1972)
121. Vance, R. V., Rock, A. O., Carr, P. W.: U.S. Patent 3654081 (1972)
122. Dale, J. K., Langlois, D. P.: U.S. Patent 2201609 (1940)
123. Pomeranz, Y., Rubenthaler, G. L., Finney, K. J.: Food Technol. **18**, 1642 (1964)
124. Tokareva, R. R., Kretovitch, V. L.: Proc. Int. Congr. Biochem., Moscow, 5th, Vol. 8. Oxford: Pergamon, 1961, p. 289
125. Beck, H., Johnson, J. A., Miller, B. S.: Cereal Chem. **15**, 841 (1957)
126. Hayden, K. J.: J. Sci. Food. Agr. **12**, 123 (1961)
127. Matz, S. A.: Cookie and Cracker Technology. Westport Connecticut: AVI Publ., 1968
128. Liebenon, R. C.: Corn Annual, Corn Refiners Assoc. Inc., Washington, D.C. 1973
129. Hurst, T. L., Turner, A. W.: U.S. Patent 3137639 (1964)
130. Hurst, T. L.: Canadian Patent 916077 (1972)
131. Bodnar, D. A., Hinman, C. W., Nelson, W. J.: U.S. Patent 3644126 (1972)
132. Murray, D. G., Luft, L. R.: Food Technol., Chicago **27**, 32 (1973)
133. Hirao, M., Sato, Y.: German Patent 1934651 (1970)
134. Mayashihara Co.: British Patent 1232645 (1971)
135. Meady, R. E., Armbrueter, F. C.: U.S. Patent 3565765 (1971)
136. Marshal, J. J., Whelan, W. J.: Chem. Ind. London **701**, 1971 (1971)
137. Tachauer, E., Cobb, J. T., Shah, Y. T.: Biotech. Bioeng. **16**, 545 (1974)
138. Closset, G. P., Cabb, J. T., Shah, Y. T.: Biotech. Bioeng. **16**, 345 (1974)
139. Subramanian, T. V.: Biotechnol. Bioeng. **18**, 1473 (1976)
140. Olson, F. O., Richardson, T.: J. Food Sci. **39**, 653 (1974)
141. Goldman, R., Goldstein, L., Katchalski, E.: In: Biochemical Aspects of Reactions on Solid Supports. Stark, G. R. (ed.). New York: Academic Press, 1971, p. 1–78
142. Marconi, W., Gulinelli, S., Morisi, F.: In: Insolubilized Enzymes. Salmona, M., Saronio, C., Garattini, S. (eds.). New York: Raven Press, 1974, p. 51–63
143. Weetall, H. H. (ed.): Immobilized Enzymes, Antigens, Antibodies and Peptides, Preparation and Characterization. New York: Dekker, 1975
144. Zaborsky, O. R.: Immobilized Enzymes. Cleveland Ohio: CRC Press, 1973
145. Messing, R. A. (ed.): Immobilized Enzymes for Industrial Reactors. New York: Academic Press, 1975
146. Mosbach, K. ed.: Immobilized Enzymes–Methods in Enzymology, Vol. 44. New York: Academic Press, 1976
147. Wingard, B. L. Jr., Katchalski-Katzir, E., Goldstein, L. (eds.): Immobilized Enzyme Principles. In: Applied Biochemistry and Bioengineering, Vol. 1. Academic Press 1976
148. Lilly, M. D., Dunnill, P.: Immobilized Enzyme Reactors. In: Immobilized Enzymes, Methods in Enzymology. Mosbach, K. (ed.) Vol. 44. New York: Academic Press, 1976, p. 717–738
149. Davis, J. C.: Biotech. Bioeng. **16**, 1113 (1974)
150. Wieth, W. R., Venkatasubramanian, K.: Chemtechn. **4** (7), 434 (1974)
151. Lilly, M. D., Dunnill, P.: Biotechnol. Bioeng. Symp. **3**, 221 (1972)
152. O'Neill, S. P., Dunnill, P., Lilly, M. D.: Biotech. Bioeng. **13**, 337 (1971)
153. Lilly, M. D., Sharp, A. K.: Chem. Eng. (London) **215**, CE12 (1968)
154. Weetall, H. H., Havewala, N. B.: In: Enzyme Engineering. Wingard, L. B. Jr. (ed.). New York: John Wiley, 1972, pp. 241–266
155. Lilly, M. D., Hornby, E. E., Crook, E. M.: Biochem. J. **100**, 718 (1966)

156. Tosa, T., Mori, T., Chibata, Y.: J. Ferm. Technolog. **49**, 552 (1971)
157. Barker, S. A., Somers, P. J., Epton, R.: Carbohyd. Res. **8**, 491 (1968)
158. Barker, S. A., Somers, P. J., Epton, R., McLaren, J. V.: Carbohyd. Res. **14**, 287 (1970)
159. Barker, S. A., Somers, P. J., Epton, R.: Carbohyd. Res. **14**, 323 (1970)
160. Horigome, T., Kasai, K., Okujama, T.: J. Biochem. **75**, 299 (1974)
161. Linko, Y. Y., Saarinen, P., Linko, M.: Biotech. Bioeng. **17**, 153 (1975)
162. Linko, Y. Y., Linko, M.: Proc. First. National Meet. Biophys. Biotechnol., Finland 1973, p. 154
163. Ledingham, W. M., Hornby, W. E.: FEBS Lett. **5**, 118 (1969)
164. Boundy, J. A., Smiley, K. L., Swanson, C. L., Hofreiter, B. T.: Carbohyd. Res. **48**, 239 (1976)
165. Robyt, J. F., French, D.: Arch. Biochem. Biophys. **122**, 8 (1967)
166. Manecke, G., Gunzel, G.: Makromolec. Chem. **51**, 199 (1962)
167. Manecke, G., Gunzel, G.: Pure Appl. Chem. **4**, 507 (1962)
168. Manecke, G., Forster, H. J.: Makromolec. Chem. **91**, 136 (1966)
169. Wykes, J. R., Dunnill, P., Lilly, M. D.: Biochim. Biophys. Acta **250**, 522 (1971)
170. Arnott, S., Fulmer, A., Scott, W. E., Moorhouse, I. C. M., Rees, D. A.: J. Mol. Biol. **90**, 269 (1974)
171. Vretbland, P., Axen, R.: Biotech. Bioeng. **15**, 783 (1973)
172. Axen, R., Ernback, S.: Eur. J. Biochem. **18**, 351 (1971)
173. Porath, J., Sundberg, L.: In: Protides of the Biological Fluids. Peeters, M. (ed.) Vol. 18. New York: Pergamon Press, 1971, p. 401
174. Porath, J., Axen, R.: Fourth International Fermentation Symposium, Kyoto, 1972
175. Caldwell, K. D., Axen, R., Porath, J.: Biotech. Bioeng. **17**, 613 (1975)
176. Caldwell, K. D., Axen, R., Bergwall, M., Porath, J.: Biotech. Bioeng. **18**, 1605 (1976)
177. Barker, S. A., Somers, P. J.: Carbohyd. Res. **14**, 257 (1970)
178. Barker, S. A., Somers, P. J., Epton, R.: Ger. Offen. 1953189 (1970)
179. Märtensson, K.: Biotech. Bioeng. **16**, 567 (1974)
180. Wilson, R. J. H., Lilly, M. D.: Biotech. Bioeng. **11**, 349 (1969)
181. Maeda, H., Suzuki, H.: Nippon Nogei Kagaku Kaishi **44**, 547 (1970)
182. Barker, S. A., Somers, P. J., Epton, R.: Carbohyd. Res. **9**, 257 (1969)
183. Gruesback, C., Rase, H. F.: Ind. Eng. Chem. Prod. Res. Develop. **11**, 74 (1974)
184. Beck, R. S., Rase, H. F.: Ind. Eng. Chem. Prod. Res. Develop. **12**, 260 (1973)
185. Walton, H. M., Eastman, J. E.: Biotechnol. Bioeng. **15**, 951 (1973)
186. Maeda, H., Suzuki, H., Yamanchi, A., Sakimae, A.: Biotech. Bioeng. **16**, 1517 (1974)
187. Smiley, K.: Biotech. Bioeng. **13**, 309 (1971)
188. Solomon, B., Levin, Y.: Biotech. Bioeng. **16**, 1161 (1974)
189. Research Group on Immobilized Enzymes.: Acad. Sinica Peking: Acta Microbiologica Sinica **13**, 25 (1973)
190. Caldwell, K. D., Axen, R., Bengwall, M., Porath, J.: Biotech. Bioeng. **18**, 1589 (1976)
191. Solomon, B., Levin, Y.: Biotech. Bioeng. **17**, 1323 (1975)
192. Weetall, H. H., Havewala, N. B., Garfinkel, H. M., Buehl, W. M., Baum, G.: Biotech. Bioeng. **16**, 169 (1974)
193. Weetall, H. H.: Applications of Immobilized Enzymes. In: Immobilized Enzymes for Industrial Reactors. Messing, R. A. (ed.). New York, San Francisco, London: Academic Press, 1975, pp. 201–233
194. Marsh, D. R., Lee, D. D., Tsao, G. T.: Biotech. Bioeng. **15**, 483 (1973)
195. Marsh, D. R., Tsao, G. T.: Biotech. Bioeng. **18**, 349 (1976)
196. Lee, D. D., Lee, Y. Y., Reilly, P. J., Collins, E. V., Tsao, G. T.: Biotech. Bioeng. **18**, 145 (1976)
197. Brillonet, J. M., Coulet, P. R., Gantheron, D. C.: Biotech. Bioeng. **19**, 125 (1977)
198. Ledingham, W. M., Ferreira, M. S. S.: Carbohyd. Res. **30**, 196 (1973)
199. Gray, C. J., Livingstone, C. M.: Biotech. Bioeng. **19**, 349 (1977)
200. Christison, J.: Chem. and Ind. **4**, 215 (1972)
201. Solomon, B., Levin, Y.: Biotech. Bioeng. **16**, 1393 (1974)
202. Maeda, S., Suzuki, H.: Agr. Chem. **36**, 1839 (1972)

203. German Patent 1193914 (1965)
204. Enevoldsen, B. S. J.: Inst. Brew. **76**, 546 (1970)
205. Märtensson, K., Mosbach, K.: Biotech. Bioeng. **14**, 715 (1972)
206. Märtensson, K.: Biotech. Bioeng. **16**, 579 (1974)
207. Mattiasson, B., Mosbach, K.: Biochim. Biophys. Acta **235**, 253 (1971)
208. Gestrelius, S., Mattiasson, B., Mosbach, K.: Biochim. Biophys. Acta **276**, 339 (1972)
209. Ger. Offen. 1916741 (1970)
210. Ger. Offen. 1958014 (1970)
211. Märtensson, K.: Biotech. Bioeng. **16**, 1567 (1974)

Advances in Biochemical Engineering

Editors: T.K. Ghose, N. Blakebrough
Managing Editor: A. Fiechter

Volume 1
1971. 70 figures. VII, 194 pages
ISBN 3-540-05400-6
Contents: The Nature of Fermentation Fluids. – Separation of Cells from Culture Media. – A Simplified Kinetic Approach to Cellulose-Cellulase System. – Production and Applications of Enzymes. – Overproduction of Microbial Metabolites and Enzymes Due to Alteration of Regulation. – The Production of Biomass from Hydrogen and Carbon Dioxide. – Liquid and Solid Hydrocarbons.

Volume 2
1972. 70 figures. V, 215 pages
ISBN 3-540-06017-0
Contents: Enzyme Engineering. – Mixed Microbial Populations. – Scale-Up of Biological Wastewater, Treatment Reactors. – Cellulose as a Novel Energy Source. – The Culture of Plant Cells. – Application of Computers in Biochemical Engineering.

Volume 3
1974. 119 figures. VI, 290 pages
ISBN 3-540-06546-6
Contents: Seminar on Topics of Fermentation Microbiology. – Genetic Problems of the Biosynthesis of Tetracycline Antibiotics. – Some Aspects of Basic Genetic Research on Fungi and Their Practical Implications. – Microbial Oxidation of Methane and Methanol. – Modelling and Stimulation in Biochemical Engineering. – Transient and Oscillatory States of Continuous Culture. – The Significance of Microbial Film in Fermenters. – Present State and Perspectives of Biochemical Engineering.

Volume 4: Engineering
1976. 87 figures. V, 172 pages
ISBN 3-540-07747-2
Contents: Transfer of Oxygen and Scale-Up in Submerged Aerobic Fermentation. – Microbial Flocs and Flocculation in Fermentation Process Engineering. – Analog/Hybrid Computation in Biochemical Engineering. – Preparation and Properties of Gel Entrapped Enzymes.

Volumes 5: Microbial Products
1977. 31 figures, 27 tables. VII, 145 pages
ISBN 3-540-08074-0
Contents: Production of Cellulolytic Enzymes by Fungi. – An Evaluation of Enzymatic Hydrolysis of Cellulosic Materials. – Nucleic Acid Damage in Thermal Inactivation of Vegetative Microorganisms. – Cellular and Microbial Models in the Investigation of Mammalian Metabolism. – The Characterization of Mixing Fermenters.

Volume 6: New Substrates
1977. 28 figures, 26 tables. V, 127 pages
ISBN 3-540-08363-4
Contents: The Role of Thiobacillus ferrooxidans in Hydrometallurgical Processes. – Cellulase Biosynthesis and Hydrolysis of Cellulosic Substances. – Metabolism of Methanol by Yeasts. – Control of Antibiotic Synthesis by Phosphate.

Volume 7: Biotechnology
1977. 112 figures, 14 tables. V, 150 pages
ISBN 3-540-08397-9
Contents: Bubble Column Bioreactors. Tower Bioreactors Without Mechanical Agitation. – Description and Operation of a Large-Scale, Mammalian Cell, Suspension Culture Facility. – A Complementary Approach to Scale-Up Simulation and Optimization of Microbial Processes. – The Redox Potential: Its Use and Control in Biotechnology.

Volume 8: Mass Transfer in Biotechnology
1978. 95 figures. V, 151 pages
ISBN 3-540-08557-2
Contents: Technical Aspects of the Rheological Properties of Microbial Cultures. – Application of Tower Bioreactors in Cell Mass Production. – Sorption Characteristics for Gas-Liquid Contacting in Mixing Vessels.

Volume 9: Microbial Processes
1978. 69 figures, 15 tables. V, 144 pages
ISBN 3-540-08606-4
Contents: Theory and Practice of Continuous Cultivation of Microorganisms in Industrial Alcoholic Processes. – Mechanism of Liquid Hydrocarbon Uptake by Microorganisms and Growth Kinetics. – Microbial Production of Hydrogen. – In vitro Synthesis of Enzymes. Physiological Aspects of Microbial Enzyme Production.

Springer-Verlag
Berlin Heidelberg New York

Polymers

Properties
and
Applications

Editorial Board:
H.-J. Cantow,
H.J. Harwood,
J.P. Kennedy, J. Meißner,
S. Okamura, G. Olivé,
S. Olivé

Springer-Verlag
Berlin
Heidelberg
New York

Volume 1

B. Rånby, J.F. Rabek

ESR Spectroscopy in Polymer Research

1977. 356 figures, 29 tables. XIV, 410 pages
ISBN 3-540-08151-8

The main purpose of this book is to collect the present available information on the applications of electron spin resonance (ESR) spectroscopy in polymer research. The book has been written both for those who want an introduction to this field, and for those who are already familiar with ESR and are interested in application to polymers. Therefore, the fundamental principles of ESR spectroscopy are first outlined, the experimental methods including computer applications are described in more detail, and the main emphasis is on the application of ESR methods to polymer problems. The authors hope that this book will provide a useful source of information by giving a coherent treatment and extensive references to original papers, reviews, and discussions in monographs and books. In this way we hope to encourage polymer chemists, organic chemists, biochemists, physicists, and material scientists to apply ESR methods to their research problems. (2519 references).

Volume 2

H.-H. Kausch

Polymer Fracture

1978. 161 figures. Approx. 340 pages
ISBN 3-540-08786-9

In the last fifteen years modern spectroscopical methods (ESR, IR) and conventional methods of structure research have permitted considerable progress in the investigation of deformation and fracture of polymeric materials. For the first time in western languages a unified view of the kinetic theory of polymer fracture is presented by one of the scientists contributing to its development.